ELECTRONIC
MEASUREMENTS

ELECTRONIC MEASUREMENTS

PHILIP KANTROWITZ, SC. D., P.E., F.A.E.S.

GABRIEL KOUSOUROU, B.S., M.A., P.E.
  Professor of Electrical Technology
  Queensborough Community College

LAWRENCE ZUCKER, SC. D., P.E.
  Associate Professor of Electrical Technology
  Queensborough Community College

Prentice-Hall, Inc., Englewood Cliffs, New Jersey   07632

Library of Congress Cataloging in Publication Data

KANTROWITZ, PHILIP.
   Electronic measurements.

   Includes index.
    1. Electronic measurements. 2. Electric measure-
ments. I. Kousourou, Gabriel, joint author.
II. Zucker, Lawrence, joint author. III. Title.
TK7878.K36   621.381′043   78–6140
ISBN 0–13–251769–8

10 9 8 7 6

Printed in the United States of America

PRENTICE-HALL INTERNATIONAL, INC., *London*
PRENTICE-HALL OF AUSTRALIA PTY. LIMITED, *Sydney*
PRENTICE-HALL OF CANADA, LTD., *Toronto*
PRENTICE-HALL OF INDIA PRIVATE LIMITED, *New Delhi*
PRENTICE-HALL OF JAPAN, INC., *Tokyo*
PRENTICE-HALL OF SOUTHEAST ASIA PTE. LTD., *Singapore*
WHITEHALL BOOKS LIMITED, *Wellington, New Zealand*

CONTENTS

2

DIRECT CURRENT, VOLTAGE, AND RESISTANCE MEASUREMENTS   *11*

3

ALTERNATING CURRENT MEASUREMENTS   *52*

4

POWER MEASUREMENTS   75

5

OSCILLOSCOPE MEASUREMENTS   102

8

9

10

11

12

SIGNAL GENERATION:
APPLICATIONS AND MEASUREMENTS   *321*

13

RADIO AND MICROWAVE
FREQUENCY SYSTEMS   *347*

PREFACE

This basic text has been developed as a one-term course in instrumentation to be used in an Associate in Applied Science (A.A.S.) program in a community college, or on a Bachelor of Electrical Technology (B.E.T.) level. The possible expansion of the text to two semesters can depend on the discretion of the instructor. College algebra and some trigonometry are minimum mathematics required for understanding the text material.

The philosophy behind writing such a text is to instruct students in electric and electronic measurements and to introduce the student to any additional, revelant, and important measuring techniques.* Each chapter contains a discussion of the principles of operation, some basic theory, and in some cases a description of some particular instruments. Review questions in each chapter and a glossary of terms appearing at the end of the text will help the student visualize the instrumentation problems.

This text differs from the usual instrumentation text in that primary emphasis is on the measurement rather than the instrument. If the instrument is unique or has special characteristics, then the instrument is discussed.

The material in the text first contains the limiting constraining factors such as sensitivity, reproducibility, accuracy percent error, response, range, etc. The text is also subdivided into groups of the most recent measuring techniques of the various areas in the electronic fields (i.e., meter measurements, transistor devices, digital instrumentation, high-fidelity audio systems and

---

*All symbols and abbreviations used in this text follow the standard adopted by *IEEE Spectrum*, trade publication of the Institute of Electrical and Electronics Engineers, Inc.

testing, recorders and recording systems, transducer systems, signal generation, applications and measurements, radio frequency systems and testing,
etc.). With this approach, the student can understand electronic measurements and make them meaningfully and accurately. This text can also be
used by students taking advanced electronic courses.

The authors wish to acknowledge the following for their assistance:
Jacqueline Harvey of the Journal of the Audio Engineering Society, New
York City, New York; Larry Zide, editor of Sagamore Publishing Co., Inc.,
Plainview, New York; Milton Aronson, publisher of *Medical Electronics* and
Measurements and Data, Pittsburgh; Chilton Publishing Co., Radnor, Penn-
sylvania; John Eargle, audio executive and author, Los Angeles; Dr. Stanley
Greenblatt, Microwave Senior Engineer; Al Hancox, Production Manager
of Beckman Instrument Co., Schiller Park, Chicago; R. F. Kerzman of
Gould Inc., Instrument Systems Division; Kenneth Arthur of Tektronix, Inc.,
Beaverton, Oregon; Dick Horn of Yellow Springs Instrument Co., Yellow
Springs, Ohio; L. C. Lynnworth of Panametrics Inc., Waltham, Massachu-
setts; Robert Gibson of Endevco Corp., San Juan Capistrano, California;
David Berman, Field Engineer of Tektronix, Inc., Beaverton, Oregon; My
Jung, Staff Engineer of Hewlett-Packard, Paramus, New Jersey.

The cover concept was developed by Mrs. Ruth Allen of Bayside, N.Y.;
the cover photo was provided courtesy of Hewlett-Packard, Palo Alto,
California; and the design was completed by the Prentice-Hall Art Department. The authors are grateful to one and all.

Finally, the authors would like to express their appreciation to Cary F.
Baker, Jr. and his assistant, Ms. Robin Lavelle; and Donald S. Rosanelli of
Prentice Hall Inc. Without their efforts, this book would not have gotten off
the ground floor.

The authors wish to dedicate this text to their wives and family: Alice
Kantrowitz, Eve and Harry Kousourou, and Jay and Robin Zucker.

PHILIP KANTROWITZ
GABRIEL KOUSOUROU
LAWRENCE ZUCKER

ELECTRONIC
MEASUREMENTS

1
MEASUREMENT PARAMETERS

1.1 INTRODUCTION

The principal aspects of the scientific method are accurate measurement, selective analysis, and mathematical formulation. Note that the first and most important is accurate measurement. Measurement is the process by which one can convert physical parameters to meaningful numbers. The importance of measurement is simply and eloquently expressed in the following statement by the famous physicist Lord Kelvin: "I often say that when you can measure what you are speaking about and can express it in numbers, you know something about it; when you cannot express it in numbers your knowledge is of a meager and unsatisfactory kind."

The science of measurement has three primary concerns:

1. The establishment of a system of units of measurement.
2. The design, development, and application of instruments and techniques.
3. The interpretation and analysis of the data so that meaningful information is derived.

The form and concept of measurement as we know it today are the ultimate result of a major change in human behavior patterns, which occurred early in human history. Early "civilized" man led a nomadic existence. Since land cultivation had not been introduced, his hunting and animal raising kept him in a constant search for new pastures. With the development of agriculture man began to settle and cluster together in villages. With this new life style there came new necessities. Tents were no longer satisfactory dwellings; more permanent structures had to be constructed. With the advent of these new advances in his craft and the needs arising from a new lifestyle, man required a more accurate measuring process.

1.2 MEASUREMENT PROCESS

The measuring process is one in which the property of an object or system under consideration is compared to an accepted standard unit, a standard defined for that particular property. The number of times the unit standard fits into the quantity being measured is the numerical measure. The numerical measure alone is meaningless unless followed by the unit used, since it identifies the characteristic or property measured.

1.3 FUNDAMENTAL UNITS

To measure an unknown we must have an accepted unit standard for the property that is to be assessed. Since there are virtually hundreds of different quantities that man is called upon to measure, it would seem that hundreds of different standard units would be required. Fortunately this is not the case. By choosing a small number of basic quantities as standards, we can define all the others in terms of these few. The basic units are called *fundamentals*, while all the others are called *derived*. The four fundamental quantities chosen as basic are length, mass, time, and electric current. Length and time, of course, are familiar, whereas mass and current are not quantities which are intuitively obvious—they require some description. A loose definition of mass is the quantity of matter contained in an object. The mass of an object, stationary or moving with a velocity less than that of light, is constant anywhere in the universe. Electric current is defined as the flow of electrons in a conductor. Current will be discussed in great detail in subsequent chapters. There are two rules that govern the choice of a basic unit for a physical quantity:

1. The basic unit must be defined in terms of that quantity which can be measured with the greatest accuracy by available instruments.
2. The basic unit must be reproducible in any well-equipped laboratory using materials and instruments generally found in a laboratory.

1.4 SYSTEMS OF UNITS

Once the fundamental quantities are defined and accepted they are named and a system of units is born. Through the years many systems of units have been proposed and used. In this text we shall use the two systems which now predominate, the S.I. (Système International d'Unite) and the English system. Both systems are based on the four fundamental quantities of length, mass, time, and current. Table 1.1 shows the names of these units in each system. The definitions of the units as listed here are continuously modified as new and more sophisticated measuring methods are encountered.

Table 1.1

	SI	English
Length	Meter m	Foot ft
Mass	Kilogram kg	Slug
Time	Second s	Second s
Current	Ampere A	Ampere A

S.I. System

Meter: The meter is defined as the distance between two fine scratch lines on a platinum-iridium bar kept at the International Bureau of Weights and Measures at Sevres, France. An exact copy is kept at the National Bureau of Standards in Washington. It has recently been defined as 1,650,763.73 wavelengths (in a vacuum) of the orange-red spectrum line of krypton-86.

Kilogram: The standard for the kilogram is a cylinder of platinum-iridium kept by the International Bureau of Weights and Measures in Sevres, France. A duplicate is kept in Washington at the National Bureau of Standards.

Second: The second was originally defined as 1/86,400 part of a mean solar day. The mean solar day is the average value throughout the year for the time interval between two successive noons. It is now defined as 9,192,631,770 periods of a particular radiation of cesium.

English System

Foot: The foot is now defined in terms of the meter, although originally it was one third of the standard yard. Since the meter is now known to a greater accuracy, the foot is defined as 1,200/3,937 (.3048) part of a standard meter.

Slug: The slug is defined as 14.59 kg.

Second: As defined previously.

Note: The units so far defined in both systems are mechanical; the electrical unit of current is the same in both systems, and is defined as follows:

Ampere: The ampere is the magnitude of a current flowing through each of two long parallel conductors separated by a distance 1 m that results in a force between them equal to 2×10^{-7} newtons/meter (N/m) of length.

1.5 DERIVED UNITS

After the system of units has been chosen it is then possible to elicit other necessary units based on these fundamentals, for example, a unit of length multiplied by a unit of length (m × m = m² = area). A unit of length cubed (m × m × m = m³ = volume) yields a unit of volume. A unit of velocity is obtained by dividing a unit of length by a unit of time (m/s = velocity). These new units which are formed by multiplying or dividing the fundamentals are called *derived*. Any derived unit, no matter how complex, can be traced back to the fundamentals.

Table 1.2 shows the definition and unit name of some of the most common mechanical and electrical derived units. Note that the units are treated as

Table 1.2

Quantity	Unit Name	Definition	Dimensional Analysis
Area	Square meter	m x m	m^2
Acceleration	Meter/s²		$\dfrac{m}{s^2}$
Force	Newton	mass x acceleration	$\dfrac{kg \times m}{s^2}$
Work	Joule	newton x m	$\dfrac{kg \times m^2}{s^2}$
Power	Watt	$\dfrac{Joule}{s}$	$\dfrac{kg \times m^2}{s^3}$
Charge	Coulomb	ampere x s	
Voltage	Volt	$\dfrac{Joule}{coulomb}$	$\dfrac{kg \times m^2}{s^2 \times A^2}$
Resistance	Ohm	$\dfrac{V}{A}$	$\dfrac{kg \times m^2}{s^3 \times A^2}$
Capacitance	Farad	$\dfrac{coulomb}{V}$	$\dfrac{A^2 \times s^4}{kg \times m^2}$
Inductance	Henry	$\dfrac{V \times s}{A}$	$\dfrac{kg \times m^2}{A^2 \times s^2}$
Frequency	Hertz	$\dfrac{1}{s} \times cycles$	$\dfrac{cycles}{s}$

if they were algebraic variables, that is,

$$m \times m = m^2 \quad \text{and} \quad \frac{m^3}{sec \cdot m} = \frac{m^2}{sec}$$

Once a unit has been used extensively and has become familiar there is a tendency to forget that that unit has its origins in some manipulations of the fundamentals.

1.6   MULTIPLES AND DIVISIONS OF UNITS

The units in actual use are divided into submultiples for the purpose of measuring quantities smaller than the unit itself. Further, multiples of the unit are designated and named so that measurements of quantities much larger than the unit are facilitated. Table 1.3 presents the prefixes used with the unit when the multiples and submultiples are to be designated. For example, the submultiple of the unit of resistance which is $1/1,000$ of an ohm is called a milliohm, and the multiple of the unit ohm which is equal to $10^6 \, \Omega$ is called the megohm.

Table 1.3

Prefix	Multiple or Sub-multiple	Symbol
tera	10^{12}	T
giga	10^9	G
mega	10^6	M
kilo	10^3	k
milli	10^{-3}	m
micro	10^{-6}	μ
nano	10^{-9}	n
pico or micro-micro	10^{-12}	p or $\mu\mu$

1.7   UNIT CONVERSIONS

There are times when converting from the multiples or submultiples can be confusing. A few examples will clarify the process.

1. Convert 0.035 A to milliamperes:

$$0.035 \text{ A} \times \frac{10^3 \text{ mA}}{1 \text{ A}} = 0.035 \times 10^3 = 35 \text{ mA}$$

2. Convert 0.00625 s to nanoseconds:

$$0.00625 \text{ s} \times \frac{10^9 \text{ ns}}{1 \text{ s}} = 0.00625 \times 10^9 = 62{,}500 \text{ ns}$$

3. Convert 100,000 Ω to megohms:

$$100{,}000 \text{ }\Omega \times \frac{10^{-6} \text{ M}\Omega}{1 \text{ }\Omega} = 100{,}000 \times 10^{-6} \text{ M}\Omega = 0.1 \text{ M}\Omega$$

4. Convert 475 MHz to terahertz:

$$475 \text{ MHz} \times \frac{10^{-6} \text{ THz}}{1 \text{ MHz}} = 475 \times 10^{-6} \text{ THz} = 0.000475 \text{ THz}$$

5. Convert 10 pF to microfarads:

$$10 \text{ pF} \times \frac{10^{-6} \text{ }\mu\text{F}}{1 \text{ pF}} = 10 \times 10^{-6} \text{ }\mu\text{F} = 0.00001 \text{ }\mu\text{F}$$

Again note that in every case the units are handled as algebraic variables so that they may be cancelled, yielding the unit sought at the outset.

1.8   STANDARDS

The process of measurement is focused on getting the so-called true value of the unknown quantity. Unfortunately this true value can never be determined exactly. We therefore define the true value as that value measured by the instrument of highest available accuracy. If we begin to consider levels of accuracy, we are essentially considering levels or echelons of standards since a manufacturer of measuring instruments must calibrate his instruments against some standard. This standard (called the working standard) is periodically compared to a working standard of a higher order of accuracy. This standard is also compared to a set of standards called the reference standards. Finally these reference standards if portable are compared to a set of working standards at the National Bureau. This process continues upward until we finally arrive at the national standards, which are the ultimate reference. Figure 1.1 shows the various echelons of standards from the everyday working standards to the national standards.

In general, measuring instruments are calibrated against the available working standards by the manufacturer, and the accuracy of the instruments is indicated in the specifications. For example, voltmeters are rated as accurate within plus or minus a percentage of the full scale deflection. Voltmeter accuracy is discussed at great length in Chapter 2.

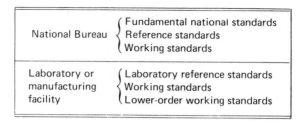

National Bureau	{ Fundamental national standards Reference standards Working standards
Laboratory or manufacturing facility	{ Laboratory reference standards Working standards Lower-order working standards

Figure 1.1

1.9   ERRORS

The error of a measurement is defined as the difference between the measured value and the true value (as defined earlier). If the reading of an ammeter is 53 mA and the value as indicated by a more accurate milliammeter is 52.1 mA, then the error is 53 mA − 52.1 mA = 0.9 mA. Errors in general are classified into four categories: human or personal, systematic, random, and applicational. In the category of human errors are those due to carelessness, inexperience, or laxity on the part of the observer. Another serious cause of human error is due to bias on the part of the individual; many times we read what we think it should be rather than what is actually indicated. Systematic or instrumental errors are due to the electrical and mechanical characteristics of the instrument. Such unavoidable problems as friction, hysteresis, or even gear backlash may be responsible for this type of error. Random errors occur in repeated measurements and are usually the most difficult to determine due to their randomness. The last and by no means the least is a condition which is encountered quite often, the misuse of an instrument—in other words, improper use of the instrument, usually contrary to the manufacturer's instructions or specifications.

1.10   ACCURACY

Accuracy is defined as the degree of agreement of the results of a measurement to the true value of the unknown. For electrical meters, the accuracy is given as a percentage of the full-scale reading; that is, if the meter has a full-scale deflection of 10 V and is said to have an accuracy of ±2%, this implies that at any setting of the meter pointer the reading can be off by 0.02 × 10 = 0.2 V. It is obvious, then, that since the error is the same regardless of the reading the absolute accuracy becomes worse as one moves away from the full-scale reading. For example, if one reads 8.6 V on this meter, the accuracy is $(0.2/8.6) \times 100 = 2.3\%$, or close to the rated accuracy. On the other hand, if one read 1.3 V, the accuracy is $(0.2/1.3) \times 100 = 15.3\%$, which is substantially worse than the specified accuracy. It is therefore important that the proper meter or range be used so that at least the specified accuracy can be achieved. In the case of multirange instruments the same accuracy applies to all ranges. It is interesting to note that although the accuracy is given as stated earlier, in essence what is really being given is the absolute error, since the percentage given is derived from dividing the error by the true value.

1.11   PRECISION

Precision is often confused with accuracy, and they should not be confused, since each means something entirely different from the other. Precision as defined in the dictionary means the quality of being exactly or sharply defined. As it pertains to the measurement field, precision means how well identically performed measurements agree with each other. A close look will reveal that there is no disagreement in the two definitions. It is literally impossible to compare a group of readings on an instrument where the scale divisions are sparse and one must estimate the next digit, whereas if an instrument has a scale with many subdivisions, the reading can be obtained more exactly to the extent that even many readings can be in total agreement. A more exactly constructed scale will therefore yield more precise readings using either of the above definitions. Note that measurements can be precise and not necessarily be accurate. Consider two clocks, one of which indicates hours and minutes, while the other indicates hours, minutes, seconds, and tenths of a second. Certainly the latter can be read more precisely, but it doesn't mean that it is indicating the right time. Being able to read the incorrect time very precisely certainly is of no use.

1.12   RANGE

The range of an instrument is defined as that region enclosed by the limits within which a particular quantity is measured. The range is usually expressed by stating the lower and upper limits. For example, the range of a voltmeter is given as 0 to 10 V, or an ammeter range is given as 0 to 200 mA. In the case of a zero-centered instrument the range is given as -10 V to $+10$ V. Where the instrument is multiranged all ranges are specified.

1.13   SPAN

The span of an instrument is the algebraic difference of the stated upper and lower limits of the range. For example, the span of the 0- to 10-V voltmeter is $10 - 0 = 10$ volt span. In the case of the zero-centered voltmeter the span is $10 - (-10) = 20$ V. For multirange instruments the span must be determined for each range.

1.14   SENSITIVITY

The sensitivity of an instrument is usually defined as the ratio of the response of the instrument to the cause or parameter being measured. Current measuring meters are usually specified as the full-scale deflection for a particular current. That is, the lowest range determines the sensitivity of the instrument. A 0- to 50-A meter is less sensitive than a meter with a 0- to 10-A range (assuming the same full-scale length). Voltmeter sensitivity is specified in ohms per volt. This specification is really an indication of the sensitivity of the meter movement (VTVM, EVM excluded). The greater the sensitivity of the meter movement, the higher the ohms-per-volt rating. Needless to say, the greater the ohms per volt, the more sensitive the meter is and the less likely it is to affect the circuit under test. The sensitivity and the range determine the input resistance of a voltmeter. Consider a voltmeter with a sensitivity of 10,000 Ω/V; the input resistance for the 0- to 10-V range is given by $10 \times 10,000 = 100,000$ Ω. Since a voltmeter is always used in parallel with potential to be measured, the resistance of the meter can affect the circuit adversely. Ideally the best voltmeter is one with infinite input resistance, but the practical meter resistance, although high in many cases, is still finite and must be considered.

1.15   LOADING EFFECT

The loading effect in measurement is that change on the circuit under test caused by the instrument used. This effect is discussed at great length as each of the measuring instruments is considered in the chapters ahead.

1.16   SUMMARY

In this chapter we have examined the process of measurement and have concluded that measurement is only a process of comparison of an unknown parameter to an accepted standard. Systems of units were considered, and the system used (S.I.) were defined. Fundamental units were contrasted to derived units, and the derivation of certain important units was shown. Multiples and divisions of units were then explored, and the conversion process from one to the other was emphasized. Finally, the various parameters of measurement such as errors, accuracy, precision, span, range, sensitivity, etc., were presented with clarifying examples of each.

REVIEW QUESTIONS

1. How does one measure? (What is the process?)
2. What is a standard, fundamental unit?
3. Why is it necessary to have divisions and multiples of the fundamentals —or for that matter of any unit?
4. Convert the following:
    a.   0.75 m to millimeters.
    b.   5 mA to amperes.
    c.   0.0001 s to nanoseconds.
    d.   0.056 MΩ to ohms.
5. Why is it important that units be standarized?
6. Give an example other than that in the chapter of the difference between precision and accuracy.
7. A thermometer has the following specification: −30°C to + 130°C. Give its range and span.
8. Which is more sensitive, a 10,000 Ω/V meter or 1,000 Ω/V meter? Explain.

2

DIRECT CURRENT, VOLTAGE, AND RESISTANCE MEASUREMENTS

2.1 INTRODUCTION

In this chapter the basic dc meter movement is introduced. Using this basic movement, we shall then discuss the design of the dc ammeter and the dc voltmeter in detail. Finally the measurement of resistance is then examined through the presentation of the various types of ohmmeters and the Wheatstone bridge. The Wheatstone bridge is again discussed in a later chapter but from a different point of view.

2.2 GALVANOMETER

A device used for the measurement or simply the detection of electric currents is called a galvanometer; it is one of the most sensitive electrical instrument and is basic to most others.

The earliest galvanometer was essentially Oersted's experimental apparatus, namely a compass needle placed below a wire in which current was to be measured. With no current flowing, both needle and wire were aligned in the north-south direction. See Fig. 2.1(a). The deflection of the compass needle when current was sent through the wire was a measure of that current. Later Lord Kelvin increased the sensitivity of this form of galvanometer (called the tangential galvanometer) by winding the wire into a coil with the compass needle in the center. See Fig 2.1(b).

Practically all galvanometers used today, however, are of the D'Arsonval

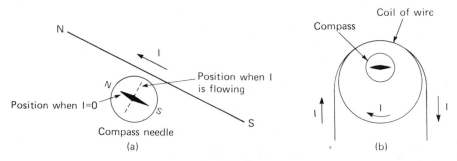

Figure 2.1

moving coil or pivoted coil types in which the roles of the magnet and the coil
are interchanged. The magnet is made larger and is shaped so that the coil,
which is now much smaller, can be suspended between its poles. The construction of a suspended moving coil galvanometer is shown in Fig. 2.2. The

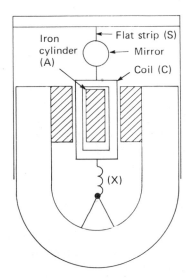

Figure 2.2

coil (C) consists of a number of turns of insulated copper wire wound on a
rectangular frame and suspended by a flat strip (S) of fine conducting wire
which provides a restoring torque when the coil is deflected from its normal
position and which also serves as a current lead to the coil. The other lead
from the coil is connected to a loosely coiled spiral (x) which is made so as
to exert a negligible force on the coil and suspension. The magnet's field is
concentrated radially by the soft iron cylinder (A) and the two curved soft

iron pole pieces of the magnet (see Fig. 2.3). When a current is passed through the coil, horizontal forces called *side thrusts* are exerted on the vertical sides of the coil due to the interaction of the current and the magnetic field.

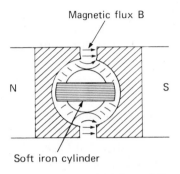

Note that the field lines of B are radially directed by the soft iron cylinder

Figure 2.3

Assume there are N turns of wire and the coil is L in. long by W in. wide (see Fig. 2.4). The force F acting perpendicular to both the direction of the current flow and the direction of the magnetic field is given as

$$F = NBIL$$

where N = number of turns,

B = magnetic field strength,

I = current,

L = vertical length of coil.

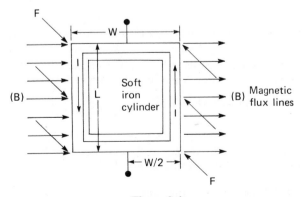

Figure 2.4

To calculate the torque we must multiply the force times the distance to the point of suspension $W/2$ so that the torque T is given by

$$T = NBIL\frac{W}{2}$$

but this is only half the torque, since there is another force being exerted on the other vertical side of the coil which is of the same magnitude but in the opposite direction. In as much as these forces form a couple we can just take the torque of one and multiply it by 2, so that the total torque is equal to

$$2T = \frac{2NBILW}{2} = NBILW = T_t, \qquad T_t = \text{total torque}$$

This torque will cause the coil to rotate until an equilibrium position is reached at an angle θ with its original orientation. In this position the torque exerted by the suspension is exactly equal to that due to the current and the field. Since the restoring torque is proportional to the displacement angle θ, we can write

$$T_t = T_s = NBILW = K\theta, \qquad T_s = \text{restoring torque}$$

where K is a constant determined by the material, the thickness, the width, and the spring constant of the suspending strip.

Solving for θ, we get

$$\theta = \frac{NBILW}{K} \quad \text{or} \quad \theta = \frac{NBLW}{K}I$$

Since $NBLW$ and K are constant for any particular galvanometer, $NBLW/K = K'$, and we get $\theta = K'I$. We now see that the angle of deflection is proportional to the current. The constant K' is called the galvanometer constant.

The angle of rotation of the coil is indicated by the use of a pointer or a reflected light beam. See Fig. 2.5.

The suspension D'Arsonval-type galvanometer can detect or measure currents as small as 10^{-12} A. The sensitivity of a galvanometer can be given in any one of three ways:

1. *Current sensitivity* (micromaperes per division): The current in microamperes required to give a deflection of one scale division.

2. *Megohm sensitivity* (ohms per volt): The resistance of the circuit (in megohms) so that the deflection will be one division with 1 V impressed:

$$\text{current sensitivity (CS)} = \frac{1}{\text{megohm sensitivity}}$$

Example

A galvanometer has a current sensitivity of 10 μA (that is, 10 μA cause a fullscale deflection). Find the megohm sensitivity.

$$\text{megohm sensitivity} = \frac{1}{\text{CS}} = \frac{1}{10 \times 10^{-6}} = 1 \times 10^5 \ \Omega/\text{V}$$

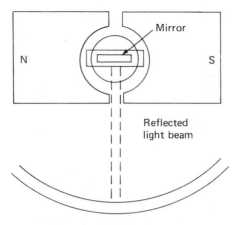

Translucent scale with
opaque calibration line

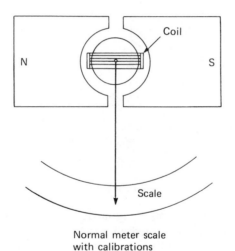

Normal meter scale
with calibrations

Figure 2.5

3. *Voltage sensitivity:* The voltage input that must be impressed on the galvanometer in series with the critical damping resistor to give a deflection of one scale division.

Although the suspension galvanometer is the most sensitive electrical instrument, it has some important disadvantages. First, due to the free vertical suspension of the coil, the instrument must be perfectly vertical. Second, it must be rigidly mounted yet free of vibrations. Third, the extreme delicacy and the required mounting exclude any possibility of the instrument being made portable. The introduction of the taut band suspension-type instrument improved portability and ruggedness but at the expense of some sensitivity.

The taut band suspension type is made so that the coil is now suspended between two strips of flat metal stretched between a mounting. See Fig. 2.6. This method eliminates any but rotational movement of the coil. The strips are also used for the restoring torque and the inducting of the current to the coil. Most portable galvanometers are of the taut band type.

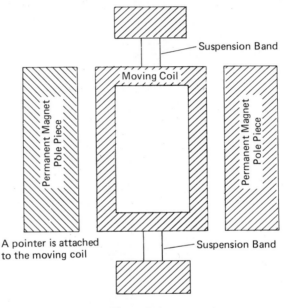

Figure 2.6

Another type of current metering movement is the pivoted coil type. In this movement the coil is mounted on a shaft through its axis, and the shaft in turn is suspended top and bottom on a jeweled bearing so that it is rigidly held with a minimum of friction. The restoring torque is applied by a coiled spring attached to the wire coil. See Fig. 2.7.

As a rule the galvanometer is used to detect the presence of current rather than to indicate its magnitude. Meter movements which indicate the magnitude of current through the use of a pointer and scale are called microammeters, milliammeters, or ammeters, depending on the sensitivity.

The meter movements form the basis of most of the current, voltage, and resistance measuring instruments. They are all dc measuring instruments. Figure 2.8 shows a picture of a pivoted coil meter movement, the most popular meter movement in use today.

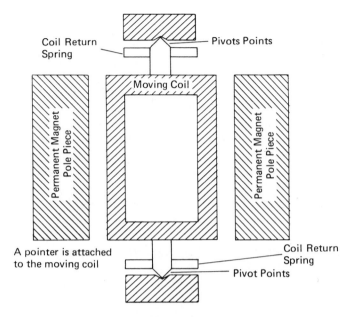

Figure 2.7

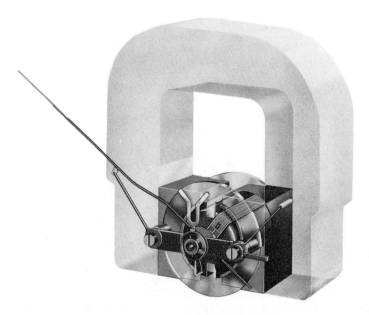

Figure 2.8 Courtesy of Weston Instruments Company, a Division of Daystrom Corp.

2.3 AMMETERS

The microammeter, as its name implies, is limited to small currents. If we wish to measure larger currents, we must adapt the meter in some way. There are two ways in which this may be done: We may (1) wind the meter coil with larger-diameter wire to increase the current-carrying capacity or (2) add some outside circuit element to increase the current-carrying capacity. Both of these methods are used in the design of milliammeters and ammeters. Milliammeters use heavier wire rather than external circuitry, while ammeters use microammeters or milliammeters with external circuitry. The external circuitry is nothing more than a resistor in parallel (shunt) with the meter movement.

In the study of parallel circuits there is the relationship called the current divider rule, which states the following: The ratio of the currents in any two parallel branches is inversely proportional to the ratio of the resistances of the branches. We can then write (see Fig. 2.9)

$$\frac{I_1}{I_2} = \frac{R_2}{R_1}$$

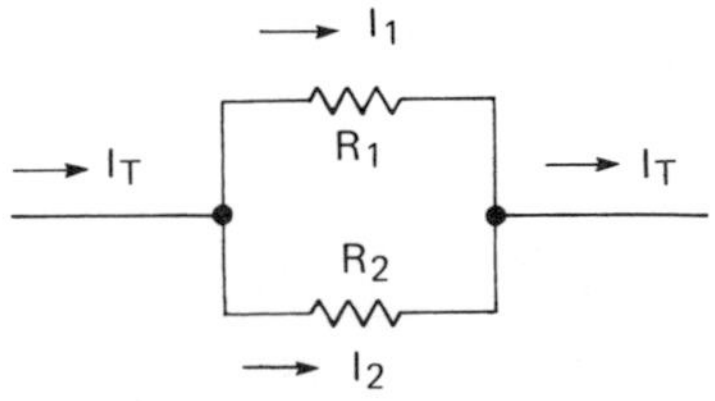

Figure 2.9

If we substitute the meter movement resistance R_m for R_1 and the shunt resistance R_s for R_2, we get (see Fig. 2.10)

$$\frac{I_m}{I_s} = \frac{R_s}{R_m} \quad \text{but} \quad I_s = I_T - I_m$$

where I_T is the new maximum current to be measured. Substituting for I_T, we get

$$\frac{I_m}{I_T - I_m} = \frac{R_s}{R_m}$$

Solving for R_m, we get

$$R_s = \frac{I_m R_m}{I_T - I_m} \tag{2.1}$$

To find the value of the shunt resistor R_s we substitute the required values into Eq. (2.1).

18

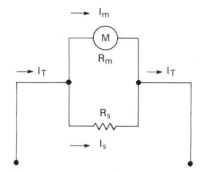

Figure 2.10

Example

Assume a meter with a coil resistance of 10 Ω and a full-scale range of 1 mA. Assume that we wish to increase the range of this instrument to 1 A full scale. We proceed as follows (see Fig. 2.11):

$$I_T = 1 \text{ A}, \qquad I_m = 0.001 \text{ A}, \qquad R_m = 10 \text{ Ω}$$

$$I_s = I_T - I_m = 1.00 - 0.001 = 0.999$$

$$R_s = \frac{0.001(10)}{0.999} = 0.01001 \text{ Ω}$$

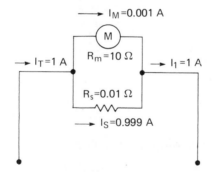

Figure 2.11

In this manner we may take any microammeter, milliammeter, or ammeter and by using suitable shunt resistors change the range. The procedure is essentially what is done in multirange ammeters. (See Fig. 2.12.) In Fig. 2.13(a) we see the front panel of a seven-range ammeter with a position marked short, which allows one to take the meter out of circuit without physically disconnecting the meter. Figure 2.13(b) shows a picture of a multirange milliammeter, the Simpson 373.

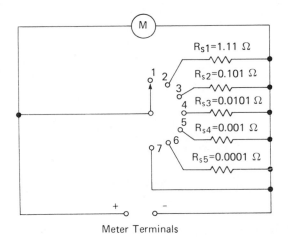

Figure 2.12

Using the meter movement of the previous example, we can calculate the resistance needed for each shunt resistor so that the meter has seven ranges as follows:

1.	0–1 mA	For 0–1 mA, no shunt resistor is needed. The value of the shunt resistor for the 0–10 mA range is calculated as follows:
2.	0–10 mA	0–10 mA, $R_{s1} = 1.11\ \Omega$.
3.	0–100 mA	$I_T = 0.01\ \text{A}, I_m = 0.001\ \text{A}, R_m = 10\ \Omega$.
4.	0–1 A	$I_s = I_T - I_m = 0.01 - 0.001 = 0.009$.
5.	0–10 A	
6.	0–100 A	
7.	Short	$R_{s1} = \dfrac{0.001 \times 10}{0.009} = 1.11\ \Omega$.

The rest of the shunt resistors are as follows:

$$R_{s2} = 0.101\ \Omega, \qquad R_{s3} = 0.0101\ \Omega, \qquad R_{s4} = 0.001\ \Omega,$$

$$R_{s5} = 0.0001\ \Omega$$

The ammeter is *always* connected in series with the load. Figures 2.14(a) and (b) indicate the proper connection of an ammeter in a circuit.

It is important, that an ammeter have as small a resistance as possible, since a large resistance would affect the operation of the circuit under test. This effect is called circuit loading. The ideal ammeter of course has zero resistance, but practically this is not realizable. The effect of loading is demonstrated in the following example.

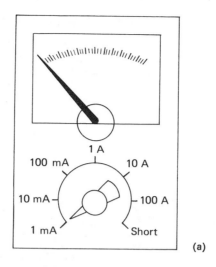

(a)

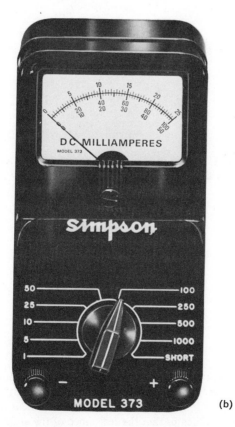

(b)

Figure 2.13 (a) and (b) Courtesy of Simpson Electric.

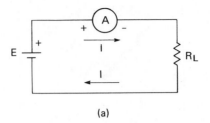

(a)

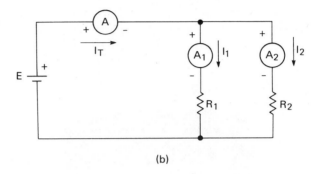

(b)

Figure 2.14

Example

Assume a circuit as shown in Fig. 2.15. By Ohm's law we compute the current $I = 100/10 = 10$ A. We wish to measure this current by inserting an ammeter

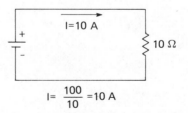

Figure 2.15

in series with the 10-Ω load. See Fig. 2.16. We read the ammeter and notice that instead of 10 A it reads 9 A. Was this the value of the original current? No, it was not. Using $I = E/R$ again, we solve for R and find that the resistance of the present circuit is 11 Ω, which indicates that the resistance of the ammeter is 1 Ω. One ohm does not seem like much resistance, but we can see that where the load is also a small resistance it can affect the circuit a great deal.

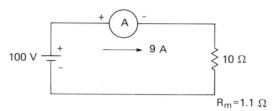

The meter reads 9 A. Therefore the total
resistance of the circuit $R_T=100/9 = 11.1\Omega$.
With the introduction of the meter the
circuit is changed. The error introduced is
% error = 10-9/10 X 100 = 10% error

Figure 2.16

Another important consideration is the observation of correct polarity. All ammeters are marked + and −, so when connecting the meter into the circuit, make sure the proper polarities are observed; otherwise the meter may be damaged. See Fig. 2.14 for the proper polarity connections.

2.4   dc VOLTMETERS

The microammeter movement which was used as the basic indicator for the ammeter may also be adapted to measure voltage. To measure voltage, a large resistor is inserted in series with the meter. This resistor limits the current through the meter so that it does not go over the full-scale rating of the meter. These resistors, whether mounted in or out of the meter case, are called multipliers.

The value of the multiplier resistor can be determined by the current for full-scale deflection and by the range of the voltage to be measured. Since the current through the meter is directly proportional to the applied voltage, from $E = R_t I_m$, where $R_t = (R_m + R_{mu})$. Assume E to be the maximum voltage to be measured.

$$R_{mu} = \frac{E}{I_m} - R_m$$

where R_m = resistance of the meter coil,

R_{mu} = multiplier resistor,

I_m = current for full-scale deflection.

Figure 2.17 shows how the weter movement is connected to operate as a voltmeter. In the actual voltmeter the meter scale is calibrated in volts.

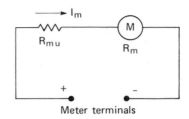

Figure 2.17

Example

We have a microammeter we wish to adapt so as to measure 1 V full scale. The meter has a resistance $R_m = 100 \, \Omega$. The full-scale current $= 100 \, \mu$A. The total resistance of the meter and the series multplier equal $R_t = R_m + R_{mu}$, but

$$R_t = \frac{E}{I_m} = \frac{1.00}{0.001} = 10{,}000 \, \Omega$$

$$\therefore 10{,}000 = 100 + R_{mu}$$

so that

$$R_{mu} = 10{,}000 - 100 = 9{,}900$$

A 9,900-Ω resistor in series with this meter will enable one to measure 1 V full scale. If we then calibrate the scale in volts, we have created a 0– to 1–V voltmeter.

As with the ammeter, multirange voltmeters can be constructed by using one meter movement and a number of multiplier resistors, one for each range. See Fig. 2.18. The voltmeter shown has six ranges: 0–10 mV,

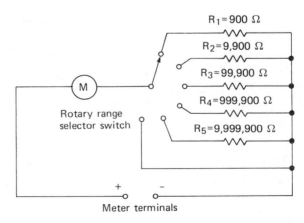

Figure 2.18

0–100 mV, 0–1 V, 0–10 V, 0–100 V, and 0–1,000 V. Using the meter from the preceding example, we can calculate each multiplier resistor as follows: For 0–10 mV:

$$R_t = \frac{E}{I} \frac{0.01}{0.0001} = 100 \ \Omega \qquad \text{(no series resistor)}$$

For 0–100 mV:

$$R_t = \frac{E}{I} = \frac{0.1}{0.0001} = 1,000 \ \Omega, \qquad R_1 = 1,000 - 100 = 900 \ \Omega$$

For 0–1 V:

$$R_t = \frac{1}{0.0001} = 10,000 \ \Omega, \qquad R_2 = 10,000 - 100 = 9,900 \ \Omega$$

For 0–10 v:

$$R_t = \frac{10}{0.0001} = 100,000 \ \Omega, \qquad R_3 = 100,000 - 100 = 99,900 \ \Omega$$

For 0–100 v:

$$R_t = \frac{100}{0.0001} = 1,000,000 \ \Omega, \qquad R_4 = 1,000,000 - 100 = 999,900 \ \Omega$$

For 0–1,000 v:

$$R_t = \frac{1,000}{0.0001} = 10,000,000 \ \Omega \qquad R_5 = 10,000,000 - 100 = 9,999,900 \ \Omega$$

The potential at any point in a circuit is always measured with reference to some other point either in or out of the circuit. If we wish to measure the voltage drop across resistor R_1 (see Fig. 2.19), we must connect the voltmeter

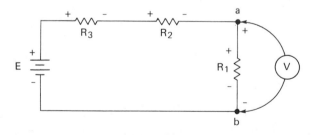

Figure 2.19

to points *a, b*. We can see that the voltmeter is across or in parallel with R_1 and the rest of the circuit. Therefore, in contrast to the ammeter, voltmeters are always connected in parallel in the circuit under test. As with the case of the ammeter, we must again observe the proper polarity when connecting a voltmeter into a circuit.

A very important consideration is the ohm-per-volt sensitivity of a volt-meter, given by

$$\text{sensitivity} = \frac{R_m + R_{mu}}{\text{F.S.}}$$

where R_m = resistance of the meter,

R_{mu} = Multiplier resistor,

F.S. = full-scale voltage value.

The input resistance of a voltmeter on any range is therefor given by

$$R_{in} = (\text{F.S.})(\text{sensitivity})$$

The input resistance of a meter of 1000 Ω/V on the 10-V range is

$$R_{in} = 10 \times 1{,}000 = 10{,}000 \ \Omega$$

The higher the sensitivity of the voltmeter, the less current will be drawn and the less circuit loading. The ideal voltmeter then would have infinite ohm-per-volt sensitivity in that the circuit parameters are not affected at all. The effect of low sensitivity loading can be demonstrated by the following example.

Example

Consider the circuit of Fig. 2.20. Suppose we wish to measure the voltage drop across R_2. The current in the circuit is 1 mA, and the voltage drop $V_{ab} = 0.001 \times 10^5 = 100$ V. We have two voltmeters available with (1) 1,000-Ω/V and (2)

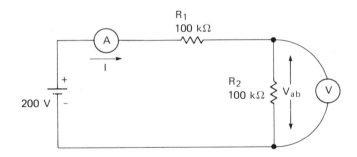

Figure 2.20

1,000,000-Ω/V sensitivities. Using voltmeter 1, we read 67 V across R_2. Checking the current again, this time with the voltmeter in the circuit, we see that the current is now 1.33 mA, so that $R_{vm} = 100$ kΩ in parallel with

$$R_2 = R_t = \frac{10^5 \times 10^5}{2 \times 10^5} = 50 \ \text{k}\Omega, \qquad V_1 = 1.33 \times 10^{-3} \times 10^5 = 133 \ \text{V}$$

By placing the voltmeter in the circuit we have changed the total resistance of the circuit from 200 kΩ to 150 kΩ, so that the original voltage drop is not there any longer. We remove voltmeter 1 and put voltmeter 2 in its place. Now we read 99 V across R_2, because the meter resistance on that range is 10^7 Ω; now 10^7 Ω in parallel with R_2 does not alter the resistance of the curcuit appreciably and does not change the potential difference being measured. This is not to say that the meter with 1,000-Ω/V sensitivity is useless but rather that the meter chosen should have a high resistance relative to the circuit under test.

There are certain types of meter movements that can measure dc as well as ac. These movements, the electrodynammeter, and the iron vane will be discussed at great length in the next chapter, in which we shall deal with the measurement of alternating current.

2.5   OHMMETERS AND RESISTANCE MEASUREMENTS

So far in this chapter we have considered the measurement of dc current and voltage by outlining the theory and operation of the more common methods and measurements. Before going on to other methods or devices used in the measurement of current and voltage we should look into the measurement of another electrical phenomenon, resistance.

Resistance as defined by Ohm's law is the constant of proportionality between the current and the voltage of a circuit. Mathematically given as $R = E/I$, quantitatively it is defined in the following way: When a current of 1 A flows through a circuit which has an impressed voltage of 1 V the circuit has a resistance of 1 Ω. This defines the unit of resistance in terms of current and voltage.

Resistance is also given in terms of the physical parameters, as indicated by the equation

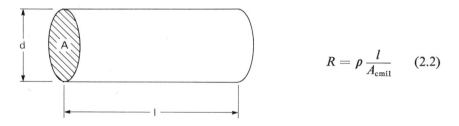

$$R = \rho \frac{l}{A_{\text{cmil}}} \qquad (2.2)$$

where R = resistance of the piece of material shown in the accompanying diagram in ohms,

I = length of the piece in feet,

A_{cmil} = area given in circular mils (cmil).

The circular mil is defined as the area of a circle whose diameter is 1 mil (1 mil = 0.001 in.). When the diameter is given in inches, convert to mils and then square to get the area in circular mils.

Example

Given the diameter $d = 0.02$ in., find the area in circular mils. Solution: Convert to mils:

$$0.02 \text{ in. } \times \frac{1,000 \text{ mils}}{\text{inch}} = 20 \text{ mils}$$

$$A_{\text{cmil}} = (20)^2 = 400 \text{ cmil}$$

ρ in Eq. (2.2) is the coefficient of resistivity. This constant is a function of the material and can be found in most handbooks of physical constants. It is always given for a particular temperature, and its units are ohm-circular mils per foot.

The following example may tend to clarify the use of Eq. (2.2).

Example

Find the resistance of a cylindrical aluminum bar that is 16 ft long with a diameter of 0.25 in. $\rho = 16.9$ at 20°C.

Convert the diameter to mils:

$$d_{\text{mils}} = 0.25 \text{ in. } \times \frac{1,000 \text{ mils}}{1 \text{ in.}} = 250 \text{ mils}$$

$$A_{\text{cmil}} = (250)^2 = 62,500 \text{ cmil}$$

$$R = \frac{16.9 \times 16}{62,500} = 0.0043 \text{ ohm}$$

$$= 4.3 \text{ milliohm}$$

There are many different ways of measuring resistance; since we can cover only a few, we shall discuss the more common ones in use today.

2.6   AMMETER-VOLTMETER METHOD

The ammeter-voltmeter method of measuring resistance is an indirect method inasmuch as we measure the current through a circuit and the voltage across it and calculate the resistance by Ohm's law. The circuit used is shown in Figs. 2.21(a) and (b).

This method is accurate when, in the circuit of Fig. 2.21(a), the resistance of the ammeter (A) is small compared to the resistance being measured. If the ammeter resistance is substantial, then this will affect the total current, and the voltmeter reading would not be of the drop across R_x but of the combination of R_x and R_A. This problem is eliminated in the circuit of Fig. 2.21(b), but another problem is added. In this circuit the current measured

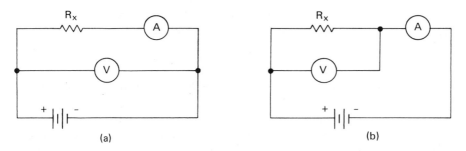

(a)   (b)

Figure 2.21

by the ammeter is actually the total of I_R and I_V. This can cause inaccuracies unless the resistance of the voltmeter V is high enough so that the current I_V is negligible.

The ohmmeter method of resistance measurement is by far the most popular and common of all the methods used. There are three types of ohmmeters, each distinguished by its internal circuitry: (1) the series type, (2) the shunt type, and (3) the voltage divider type.

2.7 SERIES OHMMETER

Figure 2.22 illustrates the circuit of the basic series ohmmeter. A battery E is connected in series with a fixed resistor R_m, a variable resistor R_2 and a 0- to 1-mA meter movement. The two leads P_1 and P_2 represent test probes which are connected across the resistance to be measured, R_x. Since the meter is rated at 0–1 mA full scale, the current in the circuit $E/R_T = 1$ mA, where $R_T = R_2 + R_m$ and E is the battery voltage. $R_2 + R_m$ must be such that when the test probes are shorted together the current through the meter is

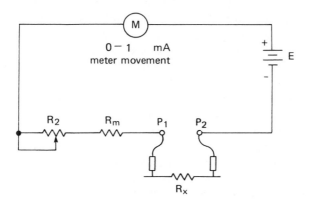

Figure 2.22

1 mA. In the circuit shown, if $E = 3$ V, then $R_2 + R_m$ must equal 3,000 Ω. $R_2 + R_m = 3{,}000$ Ω and $R_2 = 0$–500 $R_m = 2{,}700$, so that

$$\frac{E}{R_2 + R_m} = \frac{3}{3{,}000} = 0.001 = 1 \text{ mA}$$

Normally the total resistance is greater than R_T, so that the variable resistor R_2 can be used to zero the meter. R_2 is made variable so that the meter can still be zeroed when E has changed. The meter is zeroed in the following way.

The test probes are shorted together, current flows through the meter, and the pointer is deflected to the right across the scale. The scale is calibrated from right to left, with zero at the extreme right and infinity at the extreme left. An ohmmeter scale is shown in Fig. 2.24(a). Therefore, full-scale deflection indicates zero resistance across the test probes. When the test probes are shorted together the meter reads full scale or zero resistance. The variable resistor R_2 is adjusted so that the pointer reads zero. This control on the front panel of an ohmmeter is marked "zero adjust."

After the ohmmeter has been adjusted for full-scale deflection, the test probes are separated, and the meter pointer returns to the open circuit position on the left of the scale. Placing the probes across the unknown resistor connects this resistor in series with the ohmmeter circuit. As a result the current is reduced proportionately, and the meter no longer deflects full scale. If the value of R_x is equal to $R_2 + R_m = 3{,}000$, then the total resistance of the circuit becomes $3{,}000 + 3{,}000 = 6{,}000$, and the current is $3/6{,}000 = 0.5$ mA, which will deflect the pointer to the halfway mark. This point is calibrated to indicate 3,000 Ω.

2.8   SHUNT OHMMETER

The circuit of a shunt ohmmeter is shown in Fig. 2.23. When very low values of resistance are to be measured the shunt ohmmeter circuit is superior to the series ohmmeter just described. The unknown resistance R_x is now shunted across the meter, instead of in series with it. With the unknown resistor connected this way R_x acts like a shunt resistor and passes a portion of the current around the meter, and therefore the deflection of the meter drops proportionately.

In operation, the current in the ohmmeter is first adjusted by the "zero ohms adjust" R_2 to provide full-scale deflection on the meter. Contrary to the series circuit where the probes are shorted for this adjustment, in this case they are kept apart. The total resistance of the ohmmeter circuit is then adjusted so as to deflect the meter full scale. After the zero adjustment, which really does not adjust at zero but rather at the maximum resistance end of the scale, the probes are placed across R_x. Inasmuch as a portion of the meter current will now be bypassed through R_x, the meter deflection will be pro-

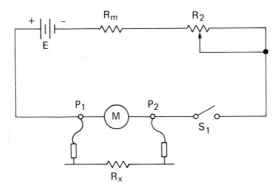

Figure 2.23

portionately less, so that minimum deflection or zero resistance results when
the total meter current is bypassed through the shorted probes.

As shown in Fig. 2.24(b), the shunt ohmmeter scale is calibrated from
left to right as opposed to the series type, which is calibrated from right to
left. Both scales are shown in Fig. 2.24.

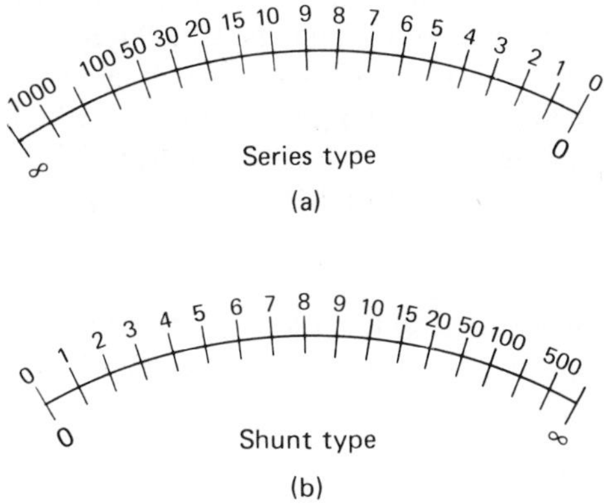

Series type

(a)

Shunt type

(b)

Note that in both cases the scale crowds up on one side.

Figure 2.24

2.9   VOLTAGE DIVIDER OHMMETER

The scale of the series ohmmeter tends to be overcrowded near the high-
resistance end of the scale. To reduce this crowding and improve the scale
division distribution, another type of circuit is used. This circuit is called the
voltage divider or potentiometer and is shown in Fig. 2.25. The meter (M)

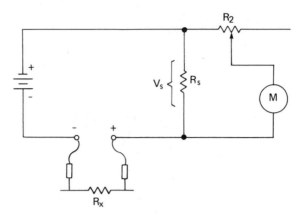

Figure 2.25

in this case is not a milliammeter but rather a high-resistance voltmeter, connected in series with R_2 and the combination across R_s. The resistor R_s and the unknown R_x form a voltage divider across the battery, so that essentially the meter reads the voltage drop across R_s. Although the meter reads volts, it is calibrated in ohms since the voltage across R_s is dependent on R_x; the larger R_x, the smaller V_s, and the smaller R_x, the larger V_s, so that when the probes are shorted together the voltage V_s is maximum (since $R_x = 0$) and the meter has a maximum deflection, which indicates that the resistance R_x is equal to zero. As with the series type, the meter reads from right to left; however, the scale divisions are somewhat more uniform than those of the series type.

An inherent disadvantage of this circuit is the fact that the battery is always in the circuit, so that unless the instrument is turned off, the battery runs down very rapidly. For this reason ohmmeters are never stored in the ohms position.

2.10 MEGGER®

When very high resistances are to be measured the currents produced with batteries are too small to be measured; it is therefore necessary to use much higher voltages. Another reason for the use of higher potential is to adequately test insulation breakdowns. This high potential is placed between the conductor and the outside surface of the insulation.

An instrument called a megger® Fig. 2.26(a), (meg ohmmeter) is used for these tests. The megger® is a portable instrument consisting of two primary elements: (1) a hand-driven dc generator G which supplies the necessary high voltage for making the test or measurement and (2) the instrument portion, which indicates the value of the resistance being measured. The instrument portion is of the opposed coil type is shown in Fig. 2.26(b). Coils *a* and *b* are mounted on the movable member (*c*) with a fixed angular relationship to each other and are free to move as a unit in the magnetic field. Coil *a* tends to move the pointer clockwise, and coil *b* tends to move the pointer counterclockwise. Coil *a* is connected in series with R_3 and R_x, the unknown resistor. The combination of R_3, R_x, and coil *a* form a direct series path between the + and − brushes of the dc generator. Coil *a* is connected in series with R_2, and these, again, are both connected across the generator. There are no restraining springs on the indicating portion of the megger®. Therefore when the generator is not operated the pointer floats freely and may come to rest at any position on the scale.

If the test leads are open-circuited, no current flows in coil *a*. However, current flows in coil *b* and deflects the pointer to infinite resistance, which indicates a resistance too large to measure. When a resistance R_x is con-

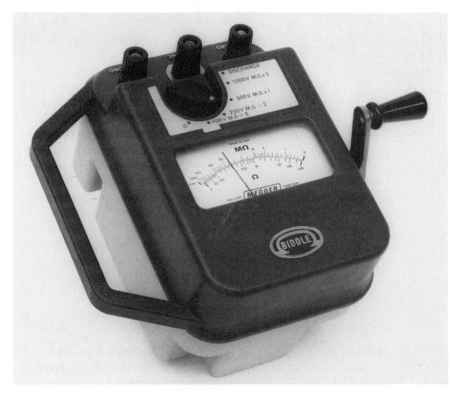

Figure 2.26 Megger® is a registered trademark of the James G. Biddle Co., Plymouth Meeting, Pennsylvania.

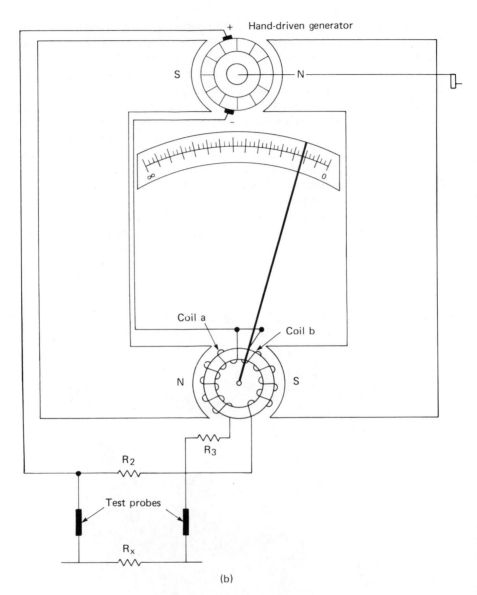

(b)

Figure 2.26   (*Continued*)

nected between the test probes, current flows through coil *a* also, tending to move the pointer clockwise. At the same time, coil *b* still tends to move the pointer counterclockwise. Therefore, the moving element (*c*) composed of both coils and the pointer comes to rest at a position at which the two forces are exactly balanced. This position depends on the value of the external resistance being measured, which controls the relative magnitude of the current in coil *a*. Because changes in voltage affect both coils *a* and *b* in the same proportion, the position of the moving system is independent of voltage. If the test leads are shorted, the pointer rests at zero because the current in coil *a* is relatively large. The instrument is not damaged due to the current in coil *a* being limited by R_3.

The external view of one type of megger® is shown in Fig. 2.26(a). Meggers® usually have a generator voltage of 500 V. To avoid excessive test voltage, most meggers® are equipped with friction clutches, so that if the generator is cranked faster than its rated speed, the clutch slips and the generator speed and output are not permitted to exceed their rated values.

2.11   WHEATSTONE BRIDGE METHOD

If we arrange four impedances in a particular configuration (Fig. 2.27) known as the Wheatstone bridge, we can derive certain interesting equations. The basic principle is as follows: If the bridge is unbalanced, there will be a current flow through *g*, the galvanometer. On the other hand, if the bridge is balanced, then the voltage V_{ab} will be equal to zero, and no current will flow through the galvanometer.

Following is the general bridge equation:

$$Z_1 Z_4 = Z_2 Z_3$$

If we construct a bridge using resistors only, then this equation would again apply. If we take three known resistors, we can find the resistance of an

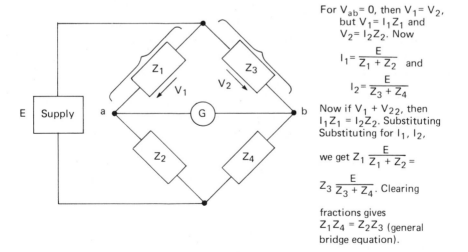

For $V_{ab} = 0$, then $V_1 = V_2$, but $V_1 = I_1 Z_1$ and $V_2 = I_2 Z_2$. Now

$$I_1 = \frac{E}{Z_1 + Z_2} \quad \text{and}$$

$$I_2 = \frac{E}{Z_3 + Z_4}$$

Now if $V_1 + V_{22}$, then $I_1 Z_1 = I_2 Z_2$. Substituting Substituting for I_1, I_2,

we get $Z_1 \dfrac{E}{Z_1 + Z_2} =$

$Z_3 \dfrac{E}{Z_3 + Z_4}$. Clearing

fractions gives $Z_1 Z_4 = Z_2 Z_3$ (general bridge equation).

Figure 2.27

unknown resistor by making the unknown the fourth arm of the bridge. See Fig. 2.28. To make the equations applicable the bridge must be balanced. To balance it R_2 is adjusted so that no current passes through the galvanometer. When there is no current the bridge is balanced and the following equation applies:

$$R_x R_4 = R_2 R_3$$

Solving for R_x, we get

$$R_x = \frac{R_3}{R_4} R_2$$

so that knowing the ratio R_3/R_4 and R_2 we can calculate R_x. With commercial Wheatstone bridges the resistance can be read right off the dials.

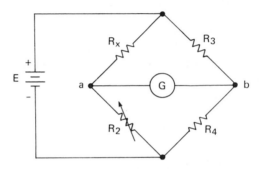

Figure 2.28

In operation there are, in the Wheatstone bridge, standard resistors in a prefixed ratio R_3/R_4. The variable resistor R_2 is a very accurate slide wire resistor with a calibrated dial. Between points a and b of the bridge a very sensitive galvanometer is connected in series with a switch, as shown in Fig. 2.29. To use the bridge, attach the unknown resistor to the terminals marked accordingly. Check to see if the correct range is selected. If the resistor to be measured is completely unknown then extreme caution must be exercised when checking the galvanometer since the currents may be in excess due to the imbalance of the bridge, and the galvanometer may be damaged. The galvanometer switch should not be depressed other than momentarily, just long enough to note if there is current flow or not. Having done this, then adjust R_2 so that the galvanometer reads zero. The setting of R_2 is noted and then multiplied by the R_3/R_4 ratio; this gives the resistance of the unknown resistor.

There are more elaborate Wheatstone bridges which are used in the laboratory. These bridges make available on the front panel the constant so that the operator may choose any ratio he may want. These resistors are normally connected by switches so that the indication on the switch is the

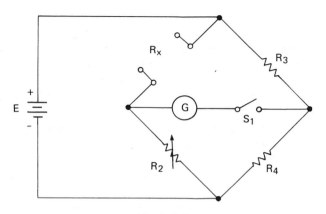

Figure 2.29

value of the resistors at that point. There is also allowance for R_2 to be varied in steps and then continuously so as to give the instrument more accuracy. Figure 2.30 shows the front panel and the schemtic of a commercial Wheatstone bridge. The bridge shown has five standard resistors for R_3, R_4, and R_2.

   Wheatstone bridges are made with ranges from 1 Ω to 1,000 MΩ. When resistances lower than 1 Ω are to be measured the Kelvin bridge is used.

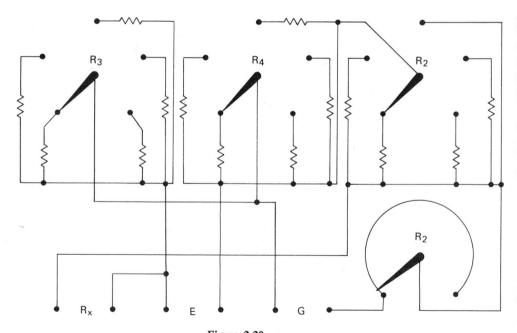

Figure 2.30

2.12 KELVIN BRIDGE METHOD

The Kelvin double bridge, or simply the Kelvin bridge, may be regarded as a modification of the Wheatstone bridge so as to improve the accuracy in the measurement of low resistances. The term double bridge is due to the presence of a double set of ratio arms. A better understanding of the Kelvin bridge may be achieved by studying the difficulties that arise in the Wheatstone bridge when measuring resistances that are so small that the resistance of the leads and contacts is appreciable. Consider the bridge shown in Fig. 2.31, where R_L represents the resistance of the lead that connects R_x to R_2.

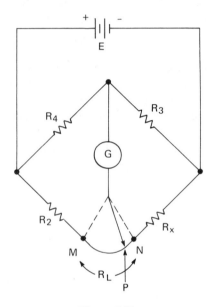

Figure 2.31

Two possible connections for the galvanometer are indicated by dotted lines. (See Fig. 2.32.) With the galvanometer at M, R_L is added to R_x, and the computed value will be higher than R_x if R_L is appreciable in comparison. On the other hand, if the connection is made to N, then R_L is added to R_2, and this will affect the balance value of that bridge arm. The computed result would then be lower than the resistor value. Suppose we can connect the galvanometer at intermediary points rather than the end points (M and N). Assume we connect it to point P, where R_L is divided into two parts such that (see Fig. 2.32)

$$\frac{R_{L_1}}{R_{L_2}} = \frac{R_3}{R_4}, \qquad R_{L_1} + R_{L_2} = R_L$$

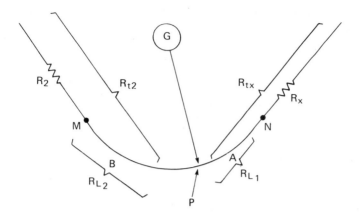

Figure 2.32

It follows from above that

$$R_{L_1} = \frac{R_3}{R_3 + R_4} R_L \quad \text{and} \quad R_{L_2} = \frac{R_4}{R_3 + R_4} R_L$$

The general bridge equation states that

$$R_{tx} = R_{t2} \frac{R_3}{R_4}$$

but

$$R_{tx} = R_x + \frac{R_3}{R_3 + R_4} R_L \quad \text{and} \quad R_{t2} = R_2 + \frac{R_4}{R_3 + R_4} R_L$$

Substituting these values into the general equation, we get

$$R_x + \frac{R_3}{R_3 + R_4} R_L = \left[R_2 + \frac{R_4}{R_3 + R_4} R_L \right] \frac{R_3}{R_4}$$

Multiplying out, we get

$$R_x + \frac{R_3}{R_3 + R_4} R_L = R_2 \frac{R_3}{R_4} + \frac{R_3}{R_3 + R_4} R_L$$

which reduces to

$$R_x = R_2 \frac{R_3}{R_4}$$

which indicates that the lead resistance is cancelled out when the galvanometer is connected to point P, and when the ratio

$$\frac{R_{L_1}}{R_{L_2}} = \frac{R_3}{R_4}$$

In practice, instead of looking for the point P so that the above ratio holds, what is done is the following: Two resistors are used, which are in the proper ratio and connected as shown in Fig. 2.33. The two resistors shown in Fig. 2.33 are connected to the galvanometer and retain the ratio

$$\frac{R'_3}{R'_4} = \frac{R_3}{R_4}$$

As shown earlier, they do not affect the measurement. They are necessary to the measurement in that they balance out the effect of the contact and lead resistance.

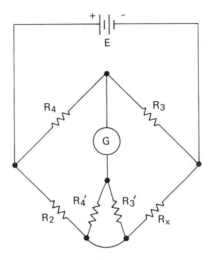

Figure 2.33

2.13 MULTIMETERS (VOMS)

In our discussion of current, voltage, and resistance measurements, we have always alluded to a single instrument designed solely for that purpose. This approach is fine where there are adequate facilities so that using a separate meter for each measurement does not present any difficulties. Consider, though, the case where these measurements must be made in the field; having to carry several instruments may prove very inconvenient. For this reason manufacturers designed the multimeter. These multimeters combine many functions; they are capable of measuring ac and dc voltage and current, resistance, and in some cases even power ratios. In fact multimeters are now more the rule than the exception. In essence the multimeter is many separate instruments in one cabinet all using the same meter movement as a readout.

To use the multimeter (VOM), one simply chooses the function and range and proceeds to measure accordingly. In Fig. 2.34 we show a typical multimeter. The one problem that is encountered is the number of scales on the meter face. See Fig. 2.35. Note that care must be exercised lest one read the

Figure 2.34 Courtesy of Simpson Electric.

wrong scale and thereby the wrong value. Although the scales are usually clearly labeled, it is still important that one read the instructional information usually accompanying the meter. To use the meter shown in Fig. 2.34, insert the test probes into the jacks labeled "common" and "+." For measuring voltages, set the mode switch to +DC and at the proper range; note that there are six ranges: 2.5 V, 10 V, 50 V, 250 V, 1,000 V, and 5,000 V. The 5,000-V range is achieved through a separate connecting jack. For measuring current again, set the mode switch to +DC and to the proper range. There are six current ranges: 1 mA, 10 mA, 100 nA, 500 mA 50 μA, and 10 A. The last two ranges have separate connecting jacks. To measure resistance the mode selector switch is placed in +DC and the function range selector is set to one of the resistance ranges; there are three: RX1, RX100,

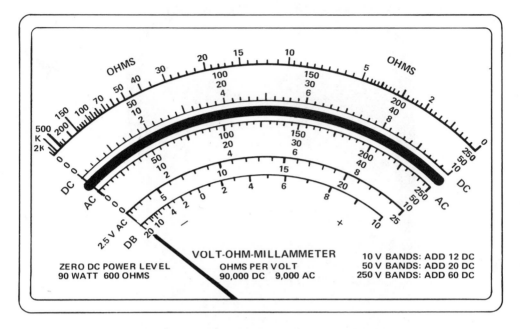

Figure 2.35   Courtesy of Simpson Electric.

and RX10,000. Note that each time switch ranges on resistance you must zero the meter; that is, short the probes together, and adjust the "zero ohms" control so that the meter reads zero. The sensitivity of the multimeters covers a wide range, from 1,000 Ω/V to 100,000 Ω/V. The sensitivity of the meter shown is 20,000 Ω/V, which is the most common. This sensitivity is usually only for the dc ranges. The actual use of the meter is the same as described for the individual instruments—voltmeters in parallel and ammeters or milli-ammeters in series. Never connect the ohmmeter to a circuit which is energized. Since multimeters contain ohmmeters, they must contain voltage sources (batteries); this fact dictates that the meter not be left on the ohm-meter function, since this would lessen the battery life. In Fig. 2.36 we show the schematic of the Simpson 260 Multimeter.

2.14   ELECTRONIC MULTIMETERS (EMMS)

The electronic multimeter originally was called the *vacuum tube voltmeter* (VTVM). With the introduction of the transistor and other semiconductor devices, vacuum tubes are no longer used in these instruments. Needless to say, the instruments are capable of measuring both dc and ac, but in this chapter we shall limit ourselves to the dc measurements only. Electronic

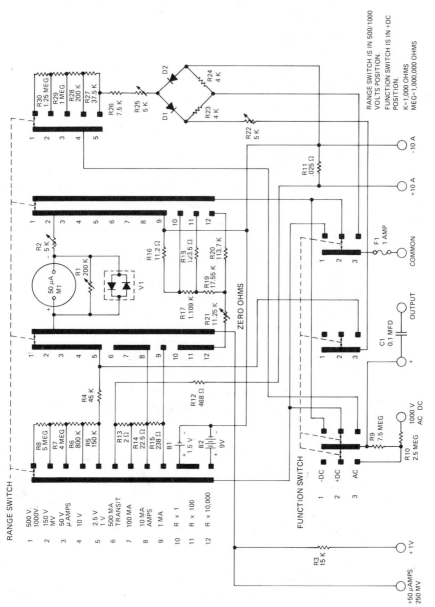

Figure 2.36 Courtesy of Simpson Electric.

43

multimeters (EMMS) fall into two categories: analog and digital. This refers to the type of readout used. When a meter movement is used as a readout the instrument is of the analog type, where as if the readout is of the various available numerical readouts, it is digital. So far the analog type is still the predominate one since it is more economical. Further, the analog meter is more suitable for special nonlinear scales such as logarithmic scales.

Although both types must perform the same function (measure voltage, current, and resistance), the principles of operation in each case are quite different.

2.15   ANALOG MULTIMETERS

Before outlining the operation of an analog electronic multimeter, let us consider the advantages and disavantages. The first important advantage is the input resistance. The input resistance of these meters is usually very high (very commonly, 11 MΩ) and is the same for all voltage ranges; this certainly alleviates the problem of loading for most voltage measuring conditions. Where the meter has current measuring functions, the resistance of the instrument is usually quite low. Here again loading effects are avoided for most current measuring situations. Resistance scales on the meter deflect in the same direction as the rest of the scales, which avoids confusion. Further, a lower voltage is used in the actual resistance measuring process so that resistors can be measured in the circuit even while in parallel with bipolar junction transistors. This feature also allows measuring junction resistance without damaging the transistor.

The basic analog electronic multimeter can be subdivided into three main blocks. See Fig. 2.37. The measuring networks are composed of various

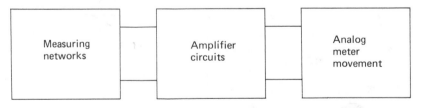

Measuring networks   Amplifier circuits   Analog meter movement

Figure 2.37

resistance networks which serve the following functions. In the case of voltage and current measurement the network is a voltage divider which limits the voltage to the amplifier and essentially sets the instrument range. Of course, we assume the voltage input is within the range of the meter, although as a rule overload protection is usually designed into the circuit. Figure 2.38 shows a typical voltage divider network. Note that in

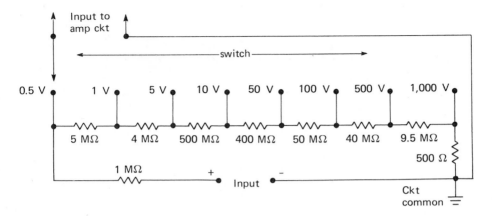

Figure 2.38

each case the voltage to the amplifier will be limited from 0 to 0.455 V for maximum input at each range. The network of Fig. 2.39 is used for current measurement. For each range starting at 1 A an additional resistor is added in series until a total of 450 Ω is inserted in the circuit at the 1-mA range. Here again we note that the voltage to the amplifier across the sensing resis-

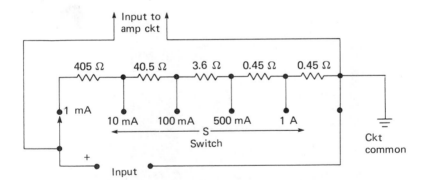

Figure 2.39

tors never exceeds 0.455 V at each range setting. The resistance measuring function is achieved through the network of Fig. 2.40. Here the unknown resistor is placed in a series circuit, and the voltage across the unknown resistor is then measured. This allows for a scale which reads from left to right, the same as the other scales in the meter. In contrast to the VOM, when measuring resistance with the electronic multimeter we must adjust both the "zero adjust" and "ohms adjust." In addition to the "zero adjust," which brings the reading to zero when the probes are shorted together, the

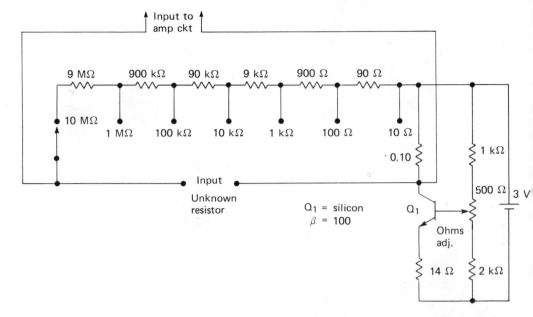

Figure 2.40

Figure 2.41 Courtesy of Hewlett-Packard Company, Palo Alto, Calif.

"ohms adjust" is used to adjust the setting of the pointer when the probes are held apart. Figure 2.41 shows an analog electronic multimeter. The amplifier block amplifies the output voltage from the measuring networks so that it can drive the analog meter movement, which is the last block.

Some of the disadvantages of this type of meter are as follows: (1) Supply voltages are required since there are amplifying circuits with active devices, and so batteries must be replaced more frequently; (2) they are more expensive to buy and to operate; and (3) the addition of the amplifying circuits introduces more components which can develop malfunctions.

2.16   DIGITAL MULTIMETERS

The digital multimeter is a comparative newcomer on the measurement scene. It is becoming increasingly more popular as the prices of these instruments become more competitive. The DMM has some important advantages. The direct numerical readout reduces the human error and the tedium involved in long periods of measurement. The increase of speed in the measurement and the elimination of the parallax error are additional advantages. Features such as autoranging and automatic polarity further reduce measurement error and any possible instrument damage through overload or reversed polarity. In some cases even hard copy in the form of printed cards or punched tape is also available. An example of a digital meter is shown in Fig. 2.42.

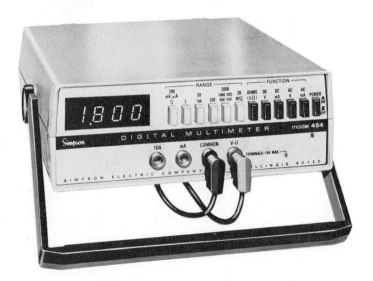

Figure 2.42   Courtesy of Simpson Electric.

Although there are numerous techniques used in DMMs, the basic principles of operation are the same. The voltage to be measured must be converted to a number of pulses proportional to the unknown voltage (V_x). The pulses are counted, and then the count is displayed on a seven-segment readout. Where the various DMMs differ is in the method used in the comparison and gating process.

In one approach the input voltage generates a pulse whose width is proportional to that voltage. The leading edge of the pulse allows the clock pulses to go to the counter, and the trailing edge stops them. The number of pulses counted will be proportional to the voltage (V_x). See Fig. 2.43.

Referring to Fig. 2.43, if $T = KV_x$, assume $V_x = 5.75$ and $K = 0.001$. Then $T = 0.001 \times 5.75 = 5.75$ ms. If the oscillator or clock has a frequency $f = 100$ KHz, Then for the period of time T, the number of pulses (n) that pass to the counter are

$$n = T \times \text{number of pulses per millisecond}$$
$$\therefore n = 0.00575 \times 10^5 = 575 \text{ pulses}$$

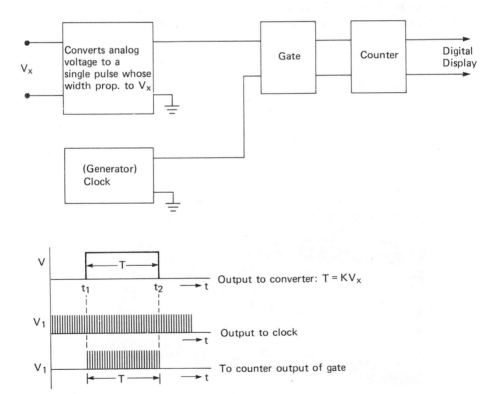

Figure 2.43

If the meter is a three-digit meter, the reading will be 5.75. The decimal point is determined by the range selector unless there is autoranging, in which case internal circuitry makes the determination.

Another method generates a ramp voltage which is compared to the input voltage V_x, and a pulse is generated whose width is proportional to the voltage V_x. The pulse, as before, is used to gate the clock pulses to the counter so that the number of pulses reaching the counter is proportional to V_x. The pulses are then counted, and the count is displayed. The process is shown in Fig. 2.44.

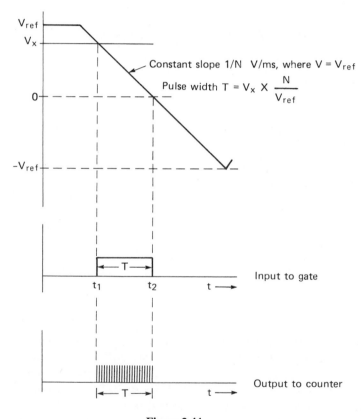

Constant slope 1/N V/ms, where $V = V_{ref}$

Pulse width $T = V_x \times \dfrac{N}{V_{ref}}$

Input to gate

Output to counter

Figure 2.44

The last technique to be discussed is probably the most preferred as of this writing. It involves the generation of a negative ramp voltage proportional to V_x. The beginning of this ramp allows clock pulses to go to the counter. When the counter reaches a predetermined number of counts then the ramp is discontinued. At this point another ramp (positive going) is generated that is proportional to an internal reference voltage until zero voltage is reached. See Fig. 2.45. During this ramp generation the counter has been counting the clock pulses which have been started at the break point. The number of pulses counted (n) are proportional to the ratio of $N(V_x/V_{ref})$, where N is the predetermined maximum count referred to earlier. The display is not functional during the generation of the negative ramp even though the counter is counting.

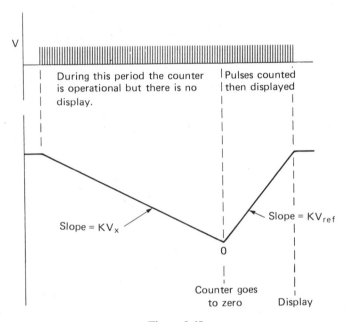

Figure 2.45

In each of the above methods it must be understood that the processes are repeated a number of times per second, so that if the input voltage is changing, this change will be indicated in the readout continuously.

The previous discussions described the operation of a digital voltmeter. Through the use of circuitry as described in the earlier Sections the DMM can be made to measure current and resistance also. The circuitry will deliver to the DMM a voltage that is proportional to the measured current or resistance, depending on the function.

Although digital voltmeters and multimeters seem to present some important advantages, they are as of now still limited in that nonlinear parameters cannot be measured. Further, for comparable accuracy they are at this time not competitively priced.

2.17 SUMMARY

In Chapter 2 we have examined the various instruments and methods used in the measurement of resistance, dc voltage, and current. We began by looking at the most fundamental of current sensitivity instruments, the galvanometer. An analysis was then made of why and how the instrument measures current. Various methods and meters for measuring voltage and current were discussed, and a general approach to voltmeter and ammeter design was indicated. Resistance measurement was then explored, looking into such instruments as series and shunt ohmmeters and resistance bridges. Finally, multimeters, electronic multimeters, and digital multimeters were discussed as to advantages and disadvantages and principles of operation.

REVIEW QUESTIONS

1. Describe a primitive current measuring device.
2. Why doesn't the suspended coil galvanometer lend itself readily to portability?
3. Show how an ammeter and voltmeter are connected to measure I_1 and V_2 in the circuit shown.

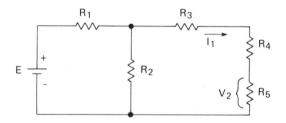

4. Given the meter current 0–10 μA and $R_m = 100\ \Omega$, design a three-range multiammeter: 0–1 mA, 0–10 mA, and 0–100 mA. Show the shunt resistors and the switching.
5. Using the above meter movement, design a three-range voltmeter: 0–1 V, 0–10 V, and 0–100 V.
6. Again using the same meter movement, design a series ohmmeter with 10 Ω, 100 Ω, and 1,000 Ω as the center scale.
7. Discuss the megger® and its uses.
8. Discuss the difference between the Wheatstone and the Kelvin bridge.
9. When using a VOM, why is it necessary to zero the meter before each resistance measurement?
10. Discuss the advantages and disadvantages of EMMs and DMMs. Compare them.

3
ALTERNATING CURRENT MEASUREMENTS

3.1 INTRODUCTION

In the previous chapter we examined the methods and instruments used in the measurement of dc voltages and currents. In this chapter we shall focus our attention on the measurement of ac voltages and currents. The measurement of ac presents some unique problems not present in the dc measurements. At the outset, ac has additional parameters besides magnitude which affect the measurement. In Fig. 3.1 we show a plot of an ac sinusoidal voltage. We shall use this as a model to define some of the basic parameters. We note that the plot is voltage versus time.

3.2 SINE WAVE

The voltage goes from zero to a positive maximum $(+E_m)$ and back to zero and then goes from zero to a maximum in the negative direction and back to zero. This complete alternation is called one cycle. The time duration for one cycle is called the period (T). The number of periodic repetitions in 1 s is called the frequency. The plot of Fig. 3.1 is called a sine wave since the instantaneous value of the voltage is given by the mathematical expression

$$e = E_m \sin (\omega t)$$

where e = instantaneous value (volts),

E_m = maximum amplitude (volts, peak value),

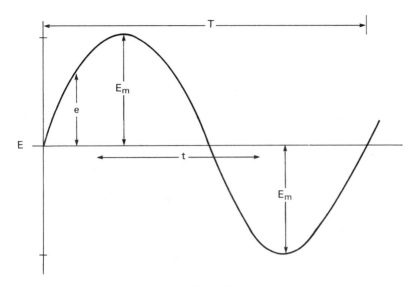

Figure 3.1

ω = radian frequency (radians per second),

t = time (seconds).

Peak-to-peak value (E_{p-p}): The peak-to-peak value is twice the maximum value. $E_{p-p} = 2E_m$.

Average value (E_{avg}): The average value of any voltage is defined mathematically as the total area under the voltage curve divided by the period in radians. In the case of the sine wave the average value is zero since the total area is zero. The positive portion is exactly equal to the negative portion; therefore the total area is zero.

Root mean square (E_{rms}): The root-mean-square or effective value of the ac is defined as that value of ac voltage which has the same heating effect as a dc voltage; that is, it requires 14.14 V of ac across a 1-Ω resistor to deliver the same heating effect as 10 V of dc. Therefore, the E_{rms} value of 14.14 V peak is 10 V, so that the ratio of

$$\frac{E_m}{E_{rms}} = \frac{14.14}{10} = 1.414 = \sqrt{2}$$

$$\therefore E_m = \sqrt{2}\, E_{rms} \quad \text{or} \quad E_{rms} = \frac{E_m}{\sqrt{2}} = 0.707E_m$$

The root-mean-square value of any periodic waveform can be calculated using the relationship

$$E_{rms} = \sqrt{\frac{1}{T} \int_0^T [e(t)]^2 \, dt}$$

In the case of the sine wave $e(t) = E_m \sin \omega t$, substituting this value into Eq. (3.1) and squaring both sides, we get

$$E_{rms}^2 = \frac{E_m^2}{T} \int_0^T \sin^2 \omega t \, dt \qquad (3.1)$$

Before integrating, we substitute the trigonometric identity

$$\sin^2 \omega t = \tfrac{1}{2} - \tfrac{1}{2} \cos 2\omega t$$

so that

$$E_{rms}^2 = \frac{E_m^2}{2T} \int_0^T (1 - \cos 2\omega t) \, dt$$

Integrating, we get

$$E_{rms}^2 = \frac{E_m^2}{2T} t \Big]_0^T - \frac{E_m^2}{4t} \sin \omega t \Big]_0^T$$

Evaluating the integrated function at $T = 2\pi$, we get

$$E_{rms}^2 = \frac{E_m^2}{2(2\pi)} 2\pi - \frac{E_m^2}{4\omega(2\pi)} (0) = \frac{E_m^2}{2} \qquad \therefore E_{rms} = \frac{E_m}{\sqrt{2}}$$

We note that the answer is the same as the one arrived at empirically for the sine wave. Of the three defined values, the rms is by far the most important, since it is the only quantity that permits a direct and accurate comparison between the effects of dc and ac signals regardless of the wave shape.

Therefore, initially we shall consider instruments which measure the rms of an ac voltage or current.

3.3 FREQUENCY RESPONSE

One very important characteristic which affects the ac measuring process is frequency. The various types of instruments used have internal resistance, capacitance, and inductance. Since the last two parameters are subject to frequency sensitivity, the frequency becomes a major influence in the measurement. The frequency spectrum is subdivided into levels or bands:

20 hz–20 kHz: audio-frequency band

10–30 kHz: very-low-frequency band

30–300 kHz: low-frequency band

300 kHz–3 MHz: medium-frequency (AM radio) band

3–30 MHz: high-frequency (CB, police) band

30–300 MHz: very-high-frequency (FM, TV) band (VHF)

300 MHz–3 GHz: ultrahigh-frequency (TV) band (UHF)

Frequencies usually above 1 GHz are said to be in the microwave range.

Our discussion of ac measurements will begin with the instruments at the low end of the spectrum.

3.4   METERS AND THE MEASUREMENT

As in the case of dc measurements the meters are inserted in the circuit— voltmeters in parallel and ammeters in series. There are certain types of measurements which might require the insertion of a capacitor in series with the voltmeter to block out any dc component in the circuit. This, of course, requires careful consideration of the effect of such capacitors on the measurement.

A knowledge of the lowest frequency that may be present in the ac to be measured is required. The value of capacitance chosen must be such that the input impedance of the meter is 100 times greater than the reactance of the capacitor at the lowest frequency present. See Fig. 3.2. Many ac instruments have a capacitor permanently connected in series with the input. It therefore is important that the specifications of the instrument be known and studied carefully.

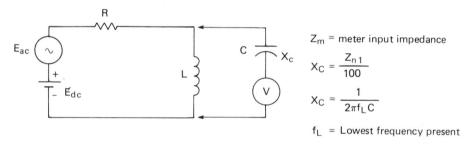

Z_m = meter input impedance

$$X_C = \frac{Z_{n1}}{100}$$

$$X_C = \frac{1}{2\pi f_L C}$$

f_L = Lowest frequency present

Figure 3.2

Meter specifications will be considered as each type of meter is introduced. Ac meters can be divided into three major categories:

1. Meters using meter movements sensitive to the rms of an ac signal.
2. Meters rectifying the ac and then using dc meter movements, calibrated in rms.
3. Meters using either a computational method or the heating value definition of rms.

There are two types of ac meter movements: the electrodynamometer type and the iron vane type.

3.5 ELECTRODYNAMOMETER MOVEMENT

The electrodynamometer is similar to the D'Arsonval dc movement except that the usual permanent magnet is replaced by a pair of stationary coils (Fig. 3.3) and the single movable coil is replaced by two coils. The fixed coils

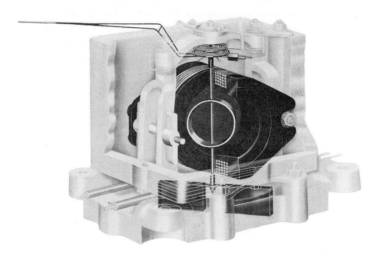

Figure 3.3   Courtesy of Weston Instruments.

are connected in series and positioned coaxially with a space between them. The two movable coils are connected in series, positioned coaxially, and pivot-mounted between the two fixed coils. The two pairs of coils (fixed and movable) are further connected in series with each other, as shown in Fig. 3.4. When a voltage is applied to the inputs A, B the same current will flow through all the coils. The interaction of the two magnetic fields will cause the pivoted coils to rotate. The force is proportional to the current in the moving coil and the magnetic field. Since the magnetic field is proportional to the same current supplied to the stationary coils, the force is proportional to the current squared (I^2). The moving parts are massive enough so that the angular position is proportional to the average force, and therefore the meter measures the average value of I^2. The scale then can be marked off to indicate the square root of the deflection. The pointer indicates the true (rms) or effective value. The circuit arrangement of an electrodynamometer-type meter is shown in Figs. 3.5(a) and (b).

Although the electrodynamometer meter is quite accurate to $\frac{1}{2}\%$ or better of FSD, it is unfortunately not very sensitive, so that it is not widely used. We shall see in the next chapter that it finds extensive use in the measurement of power. The electrodynamometer can also be used to measure dc

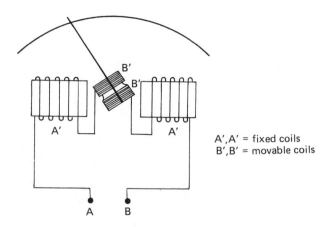

A',A' = fixed coils
B',B' = movable coils

Figure 3.4

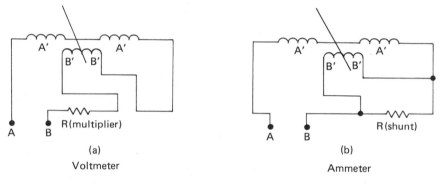

(a)
Voltmeter

(b)
Ammeter

Figure 3.5

voltages and currents since the interaction of the magnetic fields always causes an upscale deflection, no matter which way the current flows.

3.6   IRON VANE METER MOVEMENT

Another special meter movement for measuring ac is called the moving iron vane meter. The iron vane meter employs the principle of repulsion between two concentric iron vanes, one fixed and the other movable, both of which are placed inside a coil, as shown in Fig 3.6. A pointer is attached to the movable vane.

When current flows through the coil, the two iron vanes, being of a soft iron and in the magnetic flux of the coil, become magnetized. For any pa - ticular direction of current flow through the coil the vanes become magne-

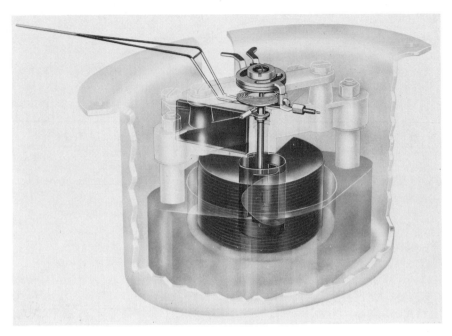

Figure 3.6   Courtesy of Weston Instruments.

tized with like poles on both ends. Since like poles repel, this force causes the movable vane to rotate against the force exerted by the restoring spring. We note from Fig. 3.6 that the movable vane is rectangular in shape, while the fixed vane is tapered. This design permits the use of a relatively uniform scale. When no current flows in the coil the movable vane is positioned so that it is opposite the larger portion of the tapered fixed vane, and the scale reading is zero. The force of repulsion is greater opposite the larger end of the fixed vane, thereby overcoming the initial inertia. The coil then rotates toward the smaller end through an angle that is propotional to the magnitude of the coil current. The movement stops when the force of repulsion is exactly equal to the restoring force of the return spring. The repulsion being always in the same direction, regardless of the direction of the current through the coil, the iron vane meter can measure ac or dc. Although this meter can be used for dc, errors are introduced due to the residual magnetism in the vanes, which makes it impractical for use in dc measurements.

Iron vane meters when used for ac measurements have an accuracy of 0.5% of FSD and better. The excellent accuracy, the fact that no current flows in the moving elements, its simplicity, and its relatively low cost make this type of movement popular in the measure of high-power, low-frequency ac voltages and currents. Due to its high magnetic reluctance, the iron vane meter requires more power than other meters and is therefore seldom used in low-power, high-resistance ac circuits. The high magnetic reluctance and large values of stray capacitance limit this meter to very low frequencies. Some laboratory instruments may have responses up to 2,500 Hz, but these are rare.

3.7 RECTIFYING AC FOR MEASUREMENT

Most of the ac measurements today are made with a dc moving coil permanent magnet movement, operating in conjunction with some kind of rectifier circuit which converts the ac to dc before it is impressed across the meter. There are two types used, the half-wave rectifier and the full-wave rectifier.

A very simple half-wave rectifier is shown in Fig. 3.7(a). The average value of the half-wave-rectified output [Fig. 3.7(b)] is given by

$$E_{avg} = \frac{E_m}{\pi} = 0.318E_m$$

Since $E_m = 1.41E_{rms}$, then $E_{avg} = 0.318(1.41)E_{rms} = 0.45E_{rms}$. Or

$$E_{rms} = \frac{E_{avg}}{0.45} = 2.22E_{avg}$$

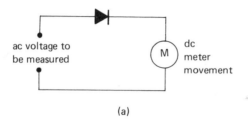

(a)

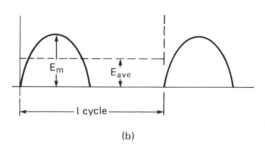

(b)

Figure 3.7

The coefficient 2.22 is called the form factor, which is the ratio of $E_{rms}/E_{avg} = 2.22$. This means that when a meter reads $E_{rms} = 100$ V it is actually measuring $100/2.22 = 45$ V; $E_{avg} = 45$ V. This factor allows us to calibrate the meter scale in E_{rms}.

When a full-wave rectifier is used [usually the bridge type; see Fig. 3.8(a) and (b)] the form factor is only 1.11 since

$$E_{\text{avg}} = \frac{2E_m}{\pi} = 0.636E_m$$

and

$$E_m = 1.41E_{\text{rms}}$$

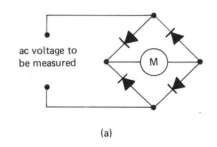

(a)

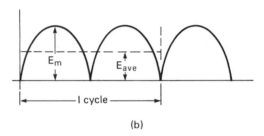

(b)

Figure 3.8

Therefore,

$$E_{\text{avg}} = 0.636(1.41)E_{\text{rms}} = 0.9E_{\text{rms}}$$

so that

$$E_{\text{rms}} = \frac{E_{\text{avg}}}{0.9} = 1.11E_{\text{avg}}$$

Here again the factor 1.11 gives the ratio of $E_{\text{rms}}/E_{\text{avg}} = 1.11$.

The actual circuit used for a half-wave rectifier meter is shown in Fig. 3.9. In this circuit an additional diode is used so that the negative half-cycle will not appear across D_2 since D_1 will conduct. By changing the value of the multiplier resistor (R_M) the range of the meter can be extended.

As indicated earlier, the rectifier-type ac voltmeter is calibrated to read the effective or rms value of the measured voltage. A meter can be modified to read peak or peak to peak by merely adding a capacitor and using the meter movement as a dc voltmeter to measure the voltage across C, the capacitor. See Fig. 3.10. Since the capacitor will charge up to the peak

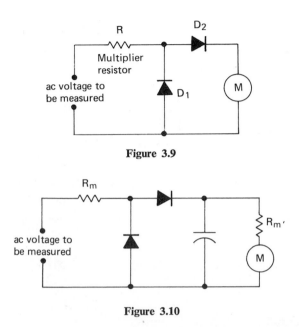

Figure 3.9

Figure 3.10

voltage, taking into consideration all the voltage division due to the various multipliers, the scale can be calibrated to read peak-to-peak values of the measured voltage. This type of meter is called peak responding and is discussed in more detail in another section of this chapter.

3.8   RECTIFIER-TYPE METERS

The rectifier-type ac meters are extensively used in applications where the accuracy requirements are not very stringent (3% to 5% FSD) and the frequency response must be well into the high-frequency band. For meters designed for a response up to 20 MHz, the rectifier is usually in the measuring probe so that a major source of pickup is eliminated. The rectifier approach is used for the ac measurement portion in most of the multimeter-type instruments. In fact, most of the electronic analog and digital meters and multimeters use the rectification method of measuring ac voltages and currents. Typical examples of a VOM and a DVM where the rectifier method is used are the Simpson Model 260 VOM and the Model 360 DVM, shown in Figs. 3.11(a) and (b). The Model 260 does not have ac current measuring capabilities, whereas the Model 360 does. In this DVM the ac voltage is measured across an accurate shunt resistor internal to the meter, the ac voltage being proportional to the current being measured. The ac voltage is measured with a half-wave operational amplifier/rectifier circuit which supplies a dc voltage and is then measured digitally by the dual slope integra-

Figure 3.11   (a) Courtesy of Simpson Electric.

tion techniques described in the previous chapter. Figure 3.12(a) shows the basic block diagram of the DVM. Figure 3.12(b) shows the ac voltage and current measuring circuits. Although average responding, the calibration gain of the circuits is based on the rms value of the sine wave.

The rectifier-type meters are usually the least expensive of all ac voltage measuring instruments.

3.9   DIRECT rms RESPONDING INSTRUMENTS

The instruments in the third category make use of a variety of methods to measure the effective value of an ac signal, each instrument using some facet of either the mathematical definition of rms or the heating effect concept.

Figure 3.11 (b) Courtesy of Simpson Electric.

The computational method relies on the actual computation of the rms value of an input voltage using circuits which perform the various mathematical operations. Figure 3.13 shows the blocks and the operation which they perform, making the overall circuit true rms responsive, independent of the wave shape. The frequency response of the computational-type instrument, although extensive (well into the high-frequency band), is still limited by the response of the circuits performing the mathematical operations. As the response of these circuits is improved, the overall response of this type of instrument will be enhanced. At this time these meters are not very common

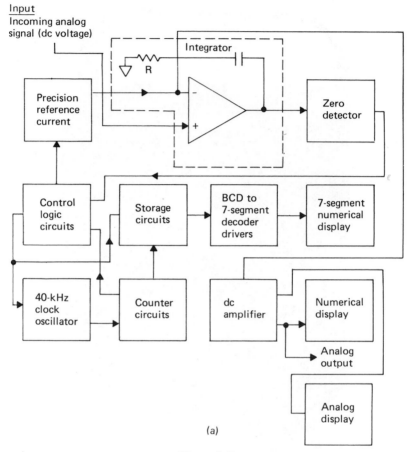

Figure 3.12

due to their high cost. Another approach which can be called true rms responsive is the measure of the heating effect of an ac signal. In this type of instrument the ac is placed across a known resistor, and the heat produced by the signals (V^2/R or I^2R for current) is measured by a thermocouple and a dc meter movement. A thermocouple is a function of two dissimilar metals across which a potential will appear, proportional to the temperature of the junction. See Fig. 3.14. Since the voltage across the thermocouple is proportional to the temperature, which is dependent on a nonlinear function (V^2/R or I^2R), the scale must also be nonlinear to the extent that it is crowded in the lower portion and progressively less crowded at the higher end. See Fig. 3.15.

Another true rms detecting meter, through the use of feedback, eliminates

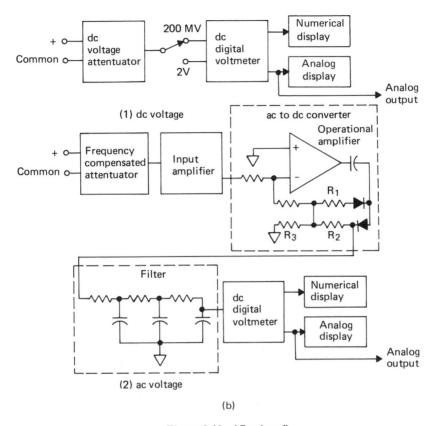

(1) dc voltage

(2) ac voltage

(b)

Figure 3.12   (*Continued*)

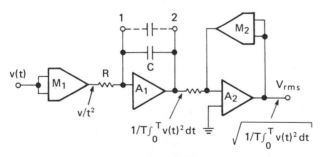

Figure 3.13

the nonlinearities introduced by the thermocouple. In this meter the detector uses two temperature measuring circuits. The voltage to the differential amplifier is proportional to the difference in temperature; this voltage is amplified and fed back to the reference resistor. When equilibrium is reached

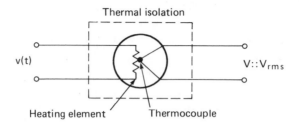

Figure 3.14

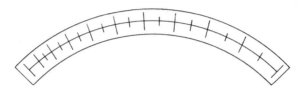

Figure 3.15

the voltage across R_f is equal to the rms value of the ac across R_{in}. See Fig. 3.16.

The true rms responding voltmeters have two important advantages. First, they have a high-frequency response, at least into the high-frequency band. Second, they are not limited by the shape of the wave. These meters will respond to the rms value of any wave shape.

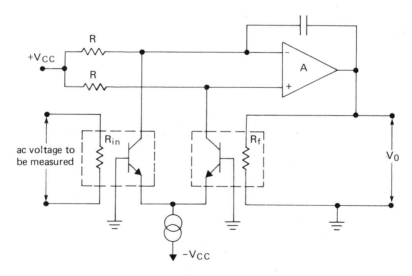

Figure 3.16

3.10 PEAK RESPONDING VOLTMETERS

In the discussion of the rectifier-type or average responding voltmeters it was pointed out that if a capacitor is placed at the output of the rectifier it would charge up to the peak value of the ac rectified signal. See Fig 3.17. If

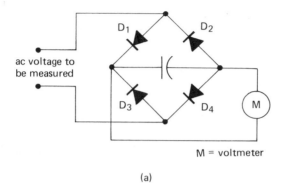

(a)

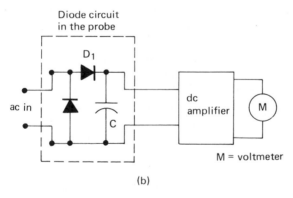

(b)

Figure 3.17

this voltage is amplified and then measured with a dc voltmeter, the meter scale can be calibrated in either peak or peak-to-peak values. The frequency response of the peak responding meter is quite high. In fact, when the peak detecting circuits are put into the probe, this type of meter may have a frequency response in excess of 1 GHz. One of the limitations of the peak detecting meter is its sensitivity, which is due to the nonlinearities of the diode detectors. In other words, when very-low-level, high-frequency signals are to be measured the diode response is so nonlinear that the measurement is very poor. This problem is eliminated through the use of a differential input, as shown in Fig. 3.18. One input of the differential amplifier is the peak value

67

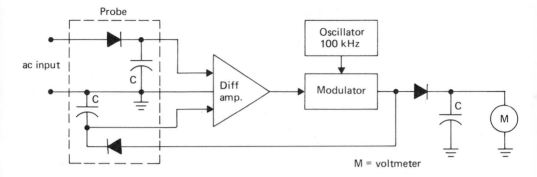

Figure 3.18

of the signal to be measured. The other input is the peak value of a rectified internally generated signal. The amplitude of this internally generated voltage is a function of the differential output of the amplifier; when the two amplitudes are the same peak value, the internal ac is equal to the unknown. Then by subjecting this voltage to another peak detector circuit and measuring it, we have the peak value of the unknown voltage.

This type of voltmeter is used where high sensitivity and high-frequency response are required. Although we indicated earlier that the actual measuring process for ac is not substantially different from that for dc, when low-level, high-frequency measurements are to be performed one must take some additional precautions to ensure accurate results. One major source of error which must be considered when low-level ac signals are being measured is that introduced by ground loops. Ground loops are present when the ground of the instrument and the ground of the device under test are at different potentials. This causes currents to flow in the ground lead, thereby introducing extraneous potentials with a resulting error in the measurement. See Fig. 3.19. In Fig. 3.19 the difference of potential between grounds is shown as an ac voltage in series with the voltage to be measured. This voltage will add to, or subtract from, the voltage E_m, thereby introducing an error. This problem can be substantially reduced by using a guard. A guarded voltmeter has the voltmeter encased in a conducting shield with an available terminal electrically connected to the guard. An instrument which has a guard will usually have four terminals for the measuring circuit. See Fig. 3.20. With the guard and the ground connected, any unwanted power-line-related signals will be shunted through the guard circuit and away from the measuring circuit.

Where the signals to be measured are low-level, very-high-frequency signals, additional precautions must be taken. At high frequencies the capacitance of the leads and the probe becomes a factor since this tends to reduce

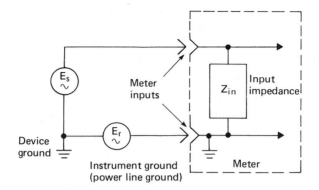

E_r = error voltage due to difference in ground potentials

Figure 3.19

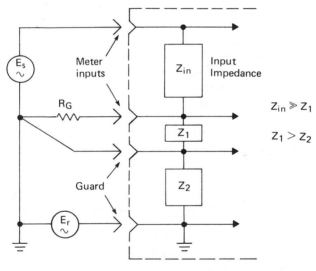

Figure 3.20

the input impedance. The reactance of 20-pF shunt capacitance at 10 MHz is

$$X_c = \frac{1}{2\pi fC} = \frac{1}{6.28 \times 10^7 \times 2 \times 10^{-11}} = \frac{1}{12.56 \times 10^{-4}} = 800 \ \Omega$$

This low reactance could present a series loading problem.

Another problem which was casually mentioned earlier but not elaborated on is the dc component that may be present with the ac signal. There are situations where the dc component, since it contributes to the rms value, must be taken into consideration in the measurement. Most ac meters are ac coupled so that the dc component is blocked out. When this is encountered one must resort to the use of an oscilloscope. Here one would measure each of the components separately and then calculate the rms value of the combination using the expression

$$V_{\text{rms(combination)}} = \sqrt{V_{\text{dc}}^2 + V_{\text{rms(ac)}}^2}$$

Most of the meters described can be used with either an analog readout or a digital readout. Whether analog or digital, the methods described in the previous chapter are used, since after the detector the voltage being measured is dc.

3.11 AC METER SPECIFICATIONS

Accuracy: The accuracy of the meter is normally given as "percent of full-scale deflection." But here one must be very careful since the accuracy will also depend on the frequency response or bandwidth of the meter. Furthermore, where the wave shape is not sinusoidal one must note whether the meter is rms responding or average responding since the readings will differ. A thorough understanding of the specifications is essential in achieving accurate results, since the manufacturer is well versed on the strong points and shortcomings of the instrument.

Input impedance: This specification is usually listed as some input resistance shunted by some capacitance. Typical values are 1–2 MΩ shunted by 20–50 pF. For some instruments (usually the rectifier type) the input impedance is given as a sensitivity, in ohms per volt, so that a meter with a 5,000-Ω/V sensitivity has an input impedance of 50,000 Ω in the 0- to 10-V range.

Frequency response (bandwidth): What is given here is the upper and lower frequency limits beyond which the stated accuracy cannot be guaranteed. The most typical meters have bandwidths from dc to 10 MHz. Certain peak responding meters have much higher upper limits.

Range: The range is as previously defined: Statement of the upper and lower limits. A meter may have as many as 20 ranges.

Sensitivity: The smallest voltage to which a meter will respond. There are ac voltmeters which have a sensitivity as low as 1 μV.

Resolution: The resolution of a meter is the degree of discernment to incremental changes in a measurement or signal level. Resolution is often expressed in parts per million and is closely related to the sensitivity. For example, a voltmeter which has a sensitivity of 10 μV should have the capability to resolve 1 μV or 1 part per million. The following specifications are exclusively given for rms responding meters.

Volt-hertz-rating: The maximum rating of the product of the input rms voltage and the frequency, consistent with the measurable frequency response and range specifications. Typical values of volt-hertz ratings are 10^5 to 10^8 V·Hz. Exceeding this specification will cause the internal amplifiers to distort the ac wave and give an inaccurate reading.

Crest factor: The ratio of peak voltage to rms value of a periodic wave is defined as the crest factor. The crest factor is a measure of the voltmeter's dynamic range:

$$\text{crest factor} = \left(\frac{1}{D-1}\right)^{1/2}$$

where D = duty cycle = ratio of one pulse width to the period of repetition. For example, a pulse train with a duty cycle of 0.015 has a crest factor of 8. Most rms responding voltmeters will handle a signal with a crest factor of 8 but may not always do so at full scale. Typical crest factor specifications range from 5 to 10.

3.12   AC CURRENT MEASUREMENTS

Alternating current measurements at frequencies below 400 cycles are usually carried out with the iron vane voltmeter. Where high currents are involved, current transformers are used extensively. Current transformers are of the voltage step-up, current step-down variety. See Fig. 3.21(a). This type of transformer is the standard closed core type which requires installation or hard wiring into the circuit. Where quick connections are to be made the split core type is more desirable. See Fig. 3.21(b). It is in fact usual to have

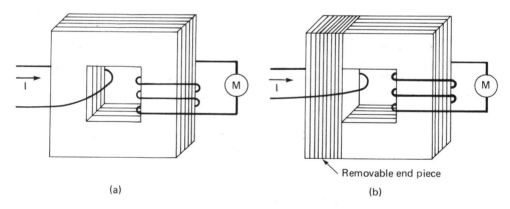

(a)   (b)

Figure 3.21

the meter attached to the transformer, so that a measurement can be made easily and expeditiously. Figure 3.22 shows this type of meter, which has come to be known as a slip-on or clamp-on ammeter. With this type of meter, it is only necessary to clamp the meter around the conductor carrying the current to be measured.

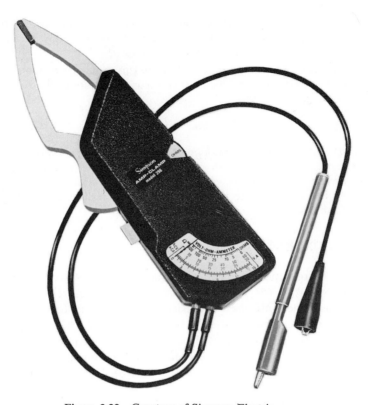

Figure 3.22   Courtesy of Simpson Electric.

For currents at frequencies above 400 cycles the average responding type of voltmeter is used in conjunction with shunt resistors. When the meter is connected in the circuit the current to be measured passes through an internal resistor of known value. The voltage drop is then measured by the rectifier-type (average responding) voltmeter whose scale is calibrated in current

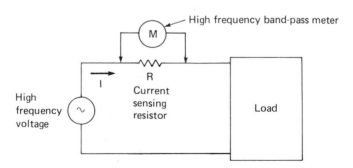

Figure 3.23

units. This type of ammeter has a bandwidth of from 1 kHz to 1 MHz depending on the design.

For higher-frequency current measurements, a high band-pass voltmeter is used in conjunction with a known carbon resistor. See Fig. 3.23. This procedure can be used well into the microwave range.

3.13 SUMMARY

Alternating current voltage and current measurements have been the theme of Chapter 3. The main emphasis was on the instruments and the factors affecting their use. The sinusoidal voltage was examined, and the important values were defined. The various levels of the frequency spectrum were specified. Three types of ac instruments and their characteristics were discussed:

1. rms responding meter movements:
   a. Not suitable for low-level measurements.
   b. Limited frequency response no higher than a few hundred cycles.
   c. High accuracy better than 1%.
   d. Medium to high price range.
2. Rectifier-type, rms calibrated instruments, usually called average responding meters:
   a. High sensitivity.
   b. Wide-band frequency response, dc–50 MHz.
   c. Accuracy usually ±3%.
   d. Least expensive of the ac measuring instruments.
   e. Not usually found as ammeters.
3. Meters using either the computational principle or the heating value definition of rms:
   a. High sensitivity.
   b. Wide-band frequency response, dc–1 GHz
   c. Some with very high accuracy better than 1%.
   d. Most expensive.

Where very high frequencies are involved the use of the peak responding meter was examined and its principle of operation analyzed.

Meter specifications such as accuracy, input impedance, frequency response, range, sensitivity resolution, volt-hertz rating, and crest factor were defined.

Finally, special ac measurement problems such as the existence of dc along with the ac signal to be measured or the introduction of error in the measurement due to ground loops were discussed.

REVIEW QUESTIONS

1. If a sinusoidal voltage is given as $e = 100 \sin \omega t$, what are the peak value (E_m), the effective value (E_{rms}), and the average value (E_{avg}) over 1 cycle?

2. Describe the principle of operation of electrodynamometer movement.

3. Give one reason the iron vane meter would naturally have a very limited frequency response.

4. A meter using a half-wave rectifier has an ac input signal equal to 100 V peak to peak. Calculate the average value of the dc being measured by the meter.

5. How would the ac portion of a signal be measured in a situation where there is an existing dc level? Explain and indicate what precautions one must take to ensure a reasonably accurate measurement.

6. Why are the rectifying circuits placed in the probe for the peak responding instrument?

7. Explain why the meter using the heating effect of the ac is impervious to frequency or wave shape.

8. Define form factor, and discuss what it means in the calibration of the instrument.

9. If a 12-V dc were applied to a full-wave rectifier ac meter, what would the meter read?

10. Under what conditions would you expect to find ground loops in a measuring circuit?

11. Why is the specification "crest factor" meaningless for sinusoidal voltages?

12. What is a current transformer, and how is it used in ac measurements?

13. Explain why a meter which measures a dc voltage proportional to the heat generated in a resistor by an ac signal has a nonlinear scale?

14. How would you measure an ac current in a circuit where the frequency was 2 GHz?

4

POWER MEASUREMENTS

4.1 POWER DEFINED

In this chapter we shall examine the measurement of both ac and dc power. Power is defined as the amount of work done in some unit of time. Mathematically,

$$\text{power } (P) = \frac{\text{work}}{\text{time}} = \frac{W}{t}$$

where work is in joules, time is in seconds, and power is in watts.

Electrical power is given as follows:

$$P = \frac{W}{t} \quad \text{but} \quad W = QV \quad \text{and} \quad t = \frac{Q}{I}$$

where $Q = $ charge in coulombs,

$V = $ potential in volts,

$I = $ current in amperes.

Substituting, we get

$$P = \frac{QV}{Q/I} = IV$$

Using Ohm's law we can derive two more relationships:

$$P = I^2 R \quad \text{and} \quad P = \frac{E^2}{R}$$

To know the power in a dc circuit, therefore, involves measuring any two of the three circuit parameters using any of the methods described in Chapter 2. This method is valid for ac circuits only when the circuit is primarily resistive. Inasmuch as this is not always true, a brief review of the factors affecting ac power may prove very helpful. Where the impressed voltage to a circuit is sinusoidal ac, reactance as well as resistance will always be found. Granted a circuit may be considered resistive where the reactances are negligible, but one must be careful to check if the measurements to be made are very sensitive. Reactance, whether capacitive or inductive, influences the current or voltage but dissipates no power. Pure reactance does not dissipate any power since it alternately stores the energy, either in the form of an electric or magnetic field, and then returns it to the circuit. In pure inductors or capacitors the power absorbed is exactly equal to the power returned, so that the net power dissipated is zero. In practice, of course, there is no such thing as a pure element, so there is always a certain amount of power dissipated. The symbols and definitions of capacitive and inductive reactance are

$$\text{inductive reactance } X_L = 2\pi f L \qquad \text{capacitive reactance } X_C = \frac{1}{2\pi f C}$$

where f is frequency in hertz, L is inductance in henries, and C is capacitance in farads.

4.2 THREE TYPES OF AC POWER

Reactances influence the current or voltage by displacing them on the time axis by 90°. The pure inductor causes the current to lag the voltage by 90°. See Fig. 4.1(a). The pure capacitor causes the current to lead the voltage by 90°. See Fig. 4.1(b). The voltage and current through a resistor are in phase. See Fig. 4.1(c). Consider the power in the circuit of Fig. 4.2. $I = I_L = I_C = I_R$. The current in a series circuit is the same through all the elements. V_L leads I by 90°, V_C lags I by 90°, and V_R is in phase with I.

Total apparent power delivered*:

$$P_A = EI \qquad \text{(volt-amperes)}$$

Reactive power delivered*:

$$P_R = EI \sin \theta \qquad \text{(vars)}$$

True average Power dissipated*:

$$P_T = EI \cos \theta \qquad \text{(watts)}$$

*In all cases E and I must be in rms units.

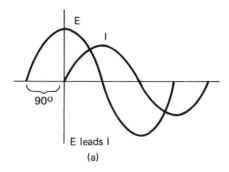

E leads I
(a)

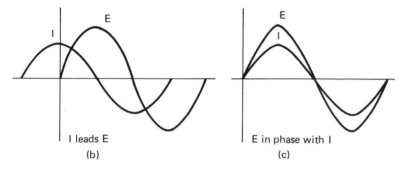

I leads E   E in phase with I
(b)   (c)

Figure 4.1   Voltage–current relationships in an ac circuit.

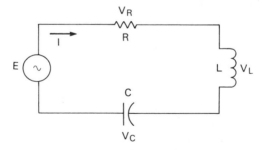

Figure 4.2   Series circuit—ac input.

where θ is the angle between the current and the voltage. The angle θ can vary from 0° to 90° depending on whether the circuit is purely resistive ($\theta = 0$) or purely reactive ($\theta = 90°$). Cos θ is called the power factor. When the circuit is resistive cos $\theta = 1$, and when it is purely reactive cos 90° = 0. Therefore the power factor (F_p) varies from 0 to 1.

The three powers are related by

$$P_A = \sqrt{P_T^2 + P_R^2}$$

where $P_R = P_{RL} - P_{RC}$,

$\quad P_{RL}$ = inductive reactive power,

$\quad P_{RC}$ = capacitive reactive power.

4.3 POWER FACTOR CORRECTION

The total power is often called the apparent power since it is not always a measure of the true power, as indicated by the preceding equation. The apparent power (P_A) is equal to the true power (P_T) only when the circuit is resistive or when $P_{RL} = P_{RC}$. In fact, when an electrical installation has a power factor of less than 1, the power company will endeavor to correct it by adding additional reactive elements to make $P_{RL} = P_{RC}$, so that it can be paid for all the power it delivers. Since power companies monitor only the true power, an installation with a power factor of less than 1 is getting more power delivered than it is paying for. Consider an installation where there are many inductive elements. With the total resistance and reactance as shown in Fig. 4.3,

$$\text{power factor } (F_p) = \cos 45 = 0.707$$

$$\text{apparent power } (P_A) = 102 \text{ V} \cdot \text{A}$$

$$\text{reactive power } (P_R) = 72 \text{ V} \cdot \text{a reactive (vars)}$$

$$\text{true power } (P_T) = 72 \text{ W}$$

The power delivered is therefore greater than the amount monitored and therefore paid for. In this case capacitors are added so that $P_{RL} = P_{RC}$ and the power delivered is equal to the true power. See Fig. 4.4.

If a capacitor of 26.5 μF is added ($X_C = 100$), the various powers are now

$$F_p = 1 \qquad P_A = 144 \text{ V} \cdot \text{A}$$

$$P_{RL} = 144 \text{ var}$$

$$P_{RC} = 144 \text{ var}$$

$$P_{RT} = 144 - 144 = 0 \text{ W}$$

$$P_T = 144 \text{ W}$$

This situation is called power factor correction, which must be used in large installations where there is a prevalence of inductive elements.

It is indicative, then, that the power to be measured is the true power inasmuch as this is the only power which is dissipated or used up. The instrument used to measure this power is called the wattmeter. The meter and measurements described in the next few sections deal with the measurement of power at very low frequencies. Power measurement at higher frequencies will be discussed later.

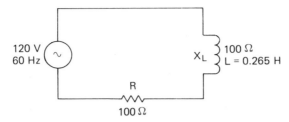

Figure 4.3   Inductive circuit.

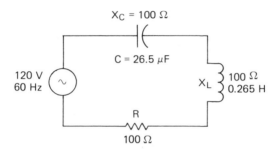

Figure 4.4   Circuit with power factor correction.

4.4   ELECTRODYNAMOMETER WATTMETER

In the last chapter the principle of the electrodynamometer-type meter movement was explained. In short there are two coils, one stationary and one
movable; the coils are connected in series, and the current to be measured is
passed through both. The deflection of the pointer is the product of the
current through each coil or the square of the current being measured. The
electrodynamometer wattmeter employs a current circuit and a potential
circuit. The current circuit consists of two fixed coils of heavy wire which are
connected in series with the line. The potential circuit consists of the two
moving coils, usually wound with much smaller-diameter wire, connected
in series with a high-valued noninductive resistance, and placed across the
line. See Figs. 4.5(a) and (b). The figure shows the internal connections of
the coils (a), the connections of the wattmeter to a load (b).

In Fig. 4.6 are shown the wattmeter connections for measuring the power
to a single load in a circuit. Notice that the current coil is in series with the
load and that the potential coil is across the load. With the load current
through the stationary coils and a current through the movable coils proportional to the load voltage there will be a torque proportional to the
product of the voltage E_L and the current I_L. When the peaks of E_L and I_C coin-
cide as in the case of a purely resistive load ($F_p = 1$) the torque is maximum.

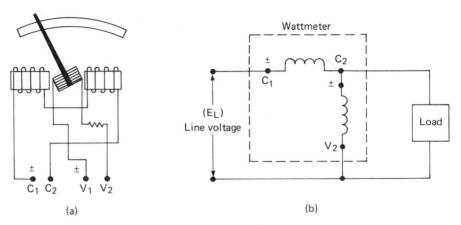

(a)   (b)

Figure 4.5   Alternating current–direct current wattmeter.

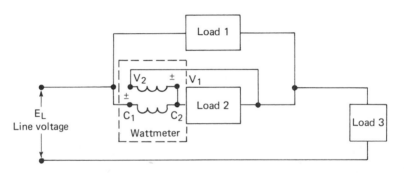

Figure 4.6   Wattmeter connected to measure power delivered to load 2.

It is minimum when the voltage and current are 90° out of phase ($F_p = 0$). It is indicative, then, that the instrument also takes phase and power factor into account.

4.5   COMPENSATED WATTMETERS

The inductance of the current coils of a wattmeter are kept as low as possible so that there is no disturbing effect on the circuit under test. The potential coil is also designed for minimum inductance so that the current through it is not only proportional to the voltage across it but also in phase with it. In fact, the noninductive resistor connected in series reduces the relative reactance to resistance ratio in the potential coil circuit so as to keep the current as closely as possible in phase with the voltage.

Since the coils of a wattmeter both have resistance, they themselves must dissipate a certain amount of power. There are wattmeters in which this is compensated for, but otherwise where the power measurement is low level, the power consumed by the instrument should be taken into consideration. The meter power loss can be easily determined by setting up the circuit of Fig. 4.7. The meter current coil is placed in series with the potential coil so that the uncompensated wattmeter reads its own power loss.

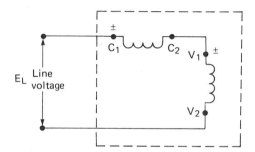

Figure 4.7   Circuit to measure power dissipated in the wattmeter.

In the compensated meter an additional coil is wound on the current coil (Fig. 4.8), and the potential coil current is made to flow opposite to the line current so that the effect of that current is cancelled out.

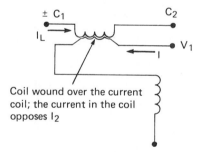

Figure 4.8   Compensated wattmeter.

Where voltmeters and ammeters are used in conjunction with a wattmeter, care must be exercised to take into consideration any and all additional power that may be consumed by the instruments. This is especially important in circuits where low levels of power are being measured.

4.6   MEASUREMENT OF POWER IN THREE-PHASE CIRCUITS

So far we have considered the measurement of power in single-phase circuits. The power in three-phase circuits can be measured in a variety of ways. Two of the most common methods are the three wattmeter and two wattmeter methods.

A load can be connected to a three-phase system in two ways, delta and wye. Figure 4.9 shows the two types of loads. Figure 4.9(a) shows a four-wire, three-phase system with a wye-connected load:

$$\text{line voltages} = E_{AB} = E_{BC} = E_{AC} = 208 \text{ V}$$

$$\text{phase voltages} = E_{AN} = E_{BN} = E_{CN} = 120 \text{ V}$$

where $E_{\text{line}} = \sqrt{3} \, E_{\text{phase}}$
Further, the phase voltages are 120° apart:

$$e_{AN} = E_{AN} \sin \omega t$$

$$e_{BN} = E_{BN} \sin (\omega t + 120°)$$

$$e_{AN} = E_{AN} \sin (\omega t + 240°)$$

In the case of the wye-connected load the phase current and line current are the same, whereas the phase and line voltages differ by the factor $\sqrt{3}$. A wye-connected load can also be attached to a three-wire system without a neutral. The phase voltages are still measured from A to N, B to N, and C to N as shown in Fig. 4.9(b). In Fig. 4.10, a delta-connected load is attached to a three-wire system. Here the phase voltage and line voltage are the same, $(E_L = E_{\text{ph}})$ but the phase currents $I_{\text{ph}(A)}$, $I_{\text{ph}(B)}$, and $I_{\text{ph}(C)}$ are related to the line currents I_{LA}, I_{LB}, and I_{LC} by the factor $\sqrt{3}$, where

$$I_L = \sqrt{3} \, I_{\text{ph}}$$

The total average or true power to a three-phase load is given by

$$P_T = 3E_{\text{ph}} I_{\text{ph}} \cos \theta$$

where θ is the phase angle between E_{ph} and I_{ph} and $\cos \theta$ is the power factor (F_p). In terms of the line voltage and current,

$$P_T = \sqrt{3} \, E_L I_L \cos \theta$$

where θ is still the angle between E_{ph} and I_{ph} and $\cos \theta$ is the power factor (F_p).

The power can be measured in two ways:

1. Three wattmeter method.
2. Two wattmeter method.

4.7   THREE WATTMETER METHOD

The connection for the measurement of the total average power to delta- or wye-connected loads is shown in Figs. 4.11(a) and (b). In each case the three wattmeter readings add up to give the total power supplied. Note that the potential coils are measuring E_{ph} and the current coils I_{ph}. Since the

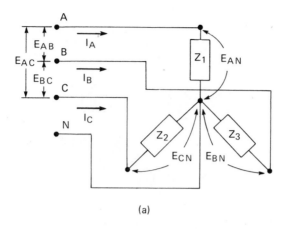

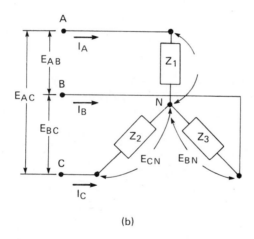

Figure 4.9 Two methods of connecting three-phase loads in a wye (y).

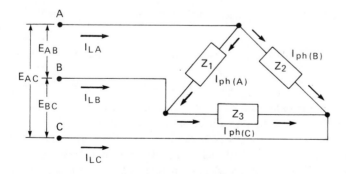

Figure 4.10 Delta (Δ)-connected three-phase circuit.

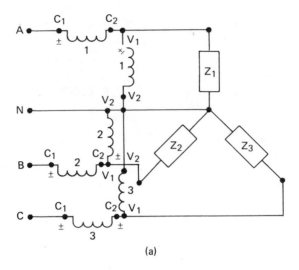

(a)

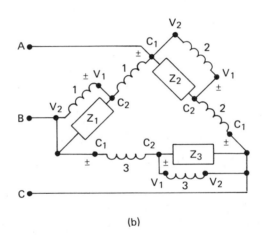

(b)

Figure 4.11   Three wattmeter method of power measurement in a three-phase circuit.

instruments are sensitive to phase, as indicated earlier, the total power will be given by

$$P_T = 3E_{ph}I_{ph}\cos\theta$$

Where the load is balanced, of course, a single wattmeter will do—merely by measuring the power to one phase and then multiplying by 3. The three wattmeter method is not always practical because very often the junction points to a delta or wye are not accessible. In fact, in a three-wire wye system the neutral is usually not accessible. In these circumstances it is more convenient to use the two wattmeter method.

4.8 TWO WATTMETER METHOD

The two wattmeter method can be used in a three-wire system, delta- or wye-connected load, balanced or unbalanced. For a four-wire system, three wattmeters are needed. The connection of the two wattmeters is shown in Fig. 4.12.

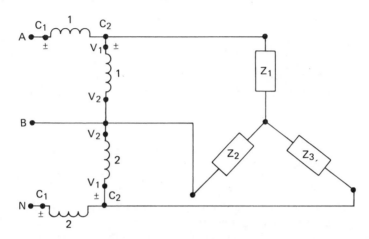

Figure 4.12 Two wattmeter method of power measurement.

It can be shown that the power read by each wattmeter is

$$P_1 = E_L I_L \cos(\theta + 30°) \qquad P_2 = E_L I_L \cos(\theta - 30°)$$

Adding the two equations, we get

$$P_1 + P_2 = E_L I_L[\cos(\theta + 30°) + \cos(\theta - 30°)] = P_T$$

To simplify and expand the bracketed terms, we substitute the identity for $\cos(A + B)$ and $\cos(A - B)$:

$$P_T = E_L I_L(\cos 30° \cos \theta - \sin 30° \sin \theta + \cos 30° \cos \theta + \sin 30° \sin \theta)$$

Simplifying by cancelling like terms, we get

$$P_T = E_L I_L(2 \cos 30° \cos \theta)$$

but $2 \cos 30° = \sqrt{3}$:

$$\therefore P_T = \sqrt{3}\ E_L I_L \cos \theta$$

which is the total power in a three-phase system. This proves that the power in a three-wire system is the the algebraic sum of the readings of the two wattmeters.

If we plot a curve of power factor (F_p) versus the ratio of two wattmeter readings P_1/P_2, where P_1 is always the smaller reading, we get a curve such as that shown in Fig. 4.13. The plot indicates that if the power factor (F_p) is

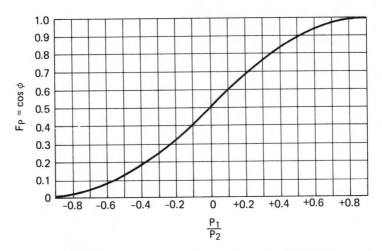

Figure 4.13 Plot of power factor (F_p) versus the ratio of P_1/P_2 in the two wattmeter method.

equal to 1 (resistive loads), then $P_1 = P_2$ and the readings are additive, $P_T = P_1 + P_2$. If the power factor is zero (reactive loads), then $P_1 = P_2$ but $P_T = 0$; therefore the readings are subtracted, $P_T = P_1 - P_2$, where the power factor is $F_p = 0.5$, One wattmeter reads zero, and the other reads the total power. That is, when $F_p > 0.5$ the readings are added, and when $F_p < 0.5$ the readings are subtracted.

Another advantage of the two wattmeter method is that it allows us to calculate the phase angle (θ) and therefore the power factor (F_p) using the relationships

$$\tan \theta = \sqrt{3}\, \frac{P_2 - P_1}{P_2 + P_1} \qquad F_p = \cos \theta \qquad .$$

This is true only where the load is balanced.

4.9 WATTMETER SPECIFICATIONS

The inexpensive wattmeters like the Simpson Model 390 have accuracies of $\pm 3\%$. There are more accurate instruments, of course, like the Weston 310, which has an accuracy of 0.25% of full-scale deflection. All will measure dc and ac power, in some cases up to 2,500 Hz. Most are multirange instruments with a high and low range depending on the voltage range used. Figure 4.14(a) shows the Simpson Model 390 ac Volt-Amp-Wattmeter with a range of 0–300–600–1,500–3,000 W. It is also a two-range ammeter and voltmeter. Figure 4.14(b) shows the Simpson Model 392 voltmeter-wattmeter. It has two ranges for voltages, 0–130 V and 0–260 V; and it has two wattage ranges, 0–1,000 W and 0–5,000 W.

(a)

(b)

Figure 4.14 Typical wattmeters. (a) Simpson model 390; (b) Simpson model 392. Courtesy of Simpson Electric.

Figure 4.15 shows two, more accurate, wattmeters: the Weston Model 310 and Model 329. Model 329 is a polyphase wattmeter consisting of two electrically independent single-phase wattmeters having their movable coil on a single shaft. Each movable coil has its own field coils. The systems can be used independently. Model 329 is made with double or triple range scales. An important consideration when using wattmeters is the current-carrying

Figure 4.15 High Accuracy wattmeter—Weston model 310. Courtesy of Weston Electrical Instrument Corp., Newark, N.J.

specification, since a meter might be used in a low-power-factor circuit, where the power reading might not exceed full scale but the current in the coils may be excessive. There are special low-power-factor wattmeters that are made to give full-scale deflection with rated volts and amperes if the power factor is 20%. Such wattmeters give better readings in low-power-factor circuits.

4.10   REACTIVE POWER MEASUREMENTS

The reactive power, as indicated earlier, is given by

$$P_R = EI \sin \theta$$

which indicates the exchange of power between the source and the load due to the reactance of the circuit.

Reactive power can be measured directly by an instrument called the varmeter. A varmeter is in effect a wattmeter in which the current in its potential coil has been shifted to lag the applied potential by 90°. Figure 4.16 shows the modified circuit of the varmeter. The current in the potential moving coil has been shifted 90° by means of a tapped inductor and a shunt resistor. The meter is now sensitive to the reactive power and is calibrated in vars.

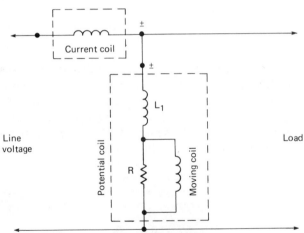

Figure 4.16   Varmeter circuit is basically a modified wattmeter.

In a three-phase, three-wire system the reactive power is given by

$$P_R = \sqrt{3} \ E_L I_L \sin \theta$$

To measure the reactive power a single-phase varmeter is connected as shown in Fig. 4.17. The meter reading must be multiplied by $\sqrt{3}$ when the meter is not calibrated for three-phase measurements.

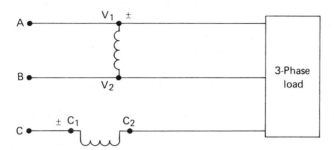

Figure 4.17 Method of measuring reactive power in a three-phase balanced circuit.

4.11 POWER-FACTOR METER

The electrodynamometer movement is extremely versatile in that it can be adapted to measure other quantities besides voltage, current, and power. If the moving coils are mounted at approximately 90° to each other, this gives rise to a new instrument, the power-factor meter. Aside from the fact that the moving coils are mounted perpendicular to each other, one coil is connected in series with a noninductive resistor and the other in series with an inductor. The current in one coil will therefore be 90° out of phase with that in the other coil. The current coils are connected in series with the line and will therefore be in phase with the line current. Figure 4.18 shows the coil arrangement of a power-factor meter. At unity power factor, the current in the moving coil in series with the resistor will be in phase with the current in the current coil. The torque produced by the interaction of their magnetic fields will cause the moving system to turn until the planes of the two coils are parallel. For this position, the pointer of the instrument would indicate a

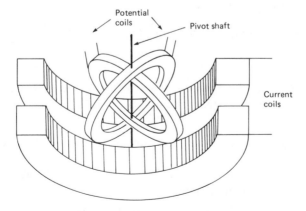

Figure 4.18 Coil arrangement of a power-factor meter.

1.00 power-factor scale marking. Since the current in the other moving coils is 90° out of phase due to the series inductor, no torque will be produced.

At zero power factor, the current in the moving coil with the series inductor will be in phase with the current in the current coil. Now the interaction of their magnetic fields will cause the moving coil to turn so that the planes of the two coils are parallel. For this position the scale marking would be 0.00 power factor. The current in the moving coil with the series resistor would be 90° out of phase with current in the circuit in the current coil and would therefore produce no torque.

For intermediate power factors, the current in each coil would have in-phase components with the current in the current coil and would produce torques in proportion to their in-phase components. The final position of the pointer would depend on the resultant torque. The power-factor meter is connected to a single-phase circuit as shown in Fig. 4.19.

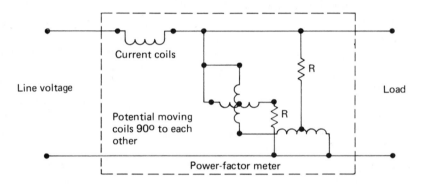

Figure 4.19   Connection of a power-factor meter to a single-phase circuit.

4.12   WATT-HOUR METER

Earlier in this chapter power was defined as

$$\text{power } (P) = \frac{\text{work}}{\text{time}} = \frac{W}{t}$$

with work in joules, time in seconds, and power in watts. If we rewrite this equation, solving for work (energy), we get

$$W = P \times t$$

so that we can express energy in watt-seconds. But the watt-second is too small a unit for practical considerations; the unit used is the kilowatthour, which is more suitable. Now 1 kw = 1,000 w and 1 h = 3,600 s; then substituting into the energy equation above, we get

$$1,000 \text{ W} \times 3,600 \text{ s} = 3.6 \times 10^6 \text{ J}$$

so that 1 kWh = 3.6×10^6 J.

The instruments which show how much energy (electrical) has been consumed are called watt-hour meters. As indicated, energy is a product of power and time; therefore the watt-hour meter must take both these factors into consideration. In principle, the watt-hour meter is a small motor whose instantaneous speed is proportional to the power passing through it. The total number of revolutions in a given time are then proportional to the total energy consumed during that time. Figures 4.20(a) and (b) show the inside construction as well as the complete watt-hour meter. Referring to Fig. 4.20(a), we see that the line is connected to the two terminals labeled source. The upper terminal S_1 is connected to two live coils F' and FF. The coil FF

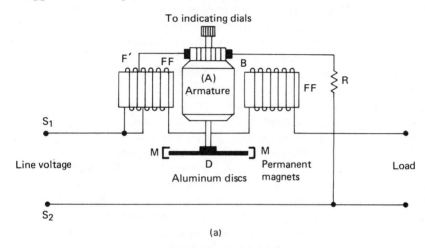

(a)

(b)

Figure 4.20   (a) Interval construction of a watt-hour meter. (b) Five dial single-phase watt-hour meter (courtesy of General Electric Corporation).

is really two coils in series as shown. They are connected in series with the
load. These coils are wound with wire sufficiently large to carry the maximum
current called for by the load. They are also connected so that their magnetic
fields are additive. The armature (A) rotates in the field produced by the
coils (FF). The remaining terminal S_2 goes directly to the load. The voltage
across the armature is the line voltage since it is connected through F' and a
small resistor R. The current in the armature is therefore proportional to the
line voltage. If the armature field is proportional to the line voltage and the
magnetic field around it is proportional to the load current, then the torque
produced must be proportional to the product of $E_L \times I_L$. In other words,
the torque is dependent on the power passing through the meter to the load.

If the meter is to register correctly, there must be a retarding torque acting
on the moving element. This force must be proportional to the speed of rotation of the moving element. To meet this condition, an aluminum disc (D) is
mounted on the armature shaft. This disc rotates between the poles of two
permanent magnets (M). In cutting through the field produced by the magnets, potentials and therefore currents are induced in the disc; the magnetic
field of these induced currents interacts with the magnetic fields of the magnets to retard the motion of the disc. The strengths of these opposing magnetic
fields are proportional to the angular velocity of the disc. Since they are interacting with a constant field strength, their retarding effect is proportional to
the speed of rotation. Coil F' is connected so that its magnetic field will add to
that of the coils (FF). This coil is added so as to overcome friction and bearing
wear. Even though the rotating elements are made as light as possible, friction
cannot be totally eliminated unless its retarding effect is counteracted by the
field of F'. The rotating element is mechanically coupled to rotating dials
which keep track of the number of revolutions and therefore the energy consumed. A typical set of watt-hour meter indicator dials is shown in Fig. 4.21.

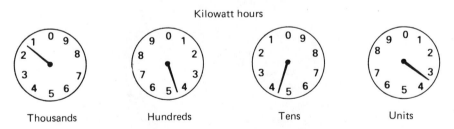

Kilowatt hours

Thousands   Hundreds   Tens   Units

Figure 4.21   Dial arrangement of a four dial watt-hour meter.

The meter illustrated in Fig. 4.21 is a four-dial instrument. The right-
hand dial registers in kilowatthours. That is, each division is 1 kWh. Each
division on the second dial is worth 10 kWh. The divisions on the third are
each 100 kWh, and finally the last has divisions each worth 1,000 kWh.

Therefore to read the dials one begins at the left and reads to the right. Let us consider the readings on the dials in Fig. 4.21. Starting at the right, we read

①   ⑤   ⑤   ③   1553 kWh

The dial hands should always be read as the figure which the pointer has last passed and not the figure it is approaching.

4.13 POWER MEASUREMENT AT HIGHER FREQUENCIES

The power measurements and instruments discussed so far have been for sources from dc to a few hundred cycles. Power measurements at higher frequencies require other types of instruments. In the audio-frequency range power measurements as a rule are made by measuring the voltage across the load resistor and calculating the power using the relationship $P = E^2/R_L$. More detailed discussions of audio power may be found in later chapters.

This method is essentially used in higher frequencies also, but it is incorporated in a single instrument called the absorption-type power meter. The absorption power meter is nothing more than a resistor, which is designed to maintain constant resistance over the range of frequencies of interest, and a high-frequency voltmeter, which is calibrated in power units. Figure 4.22 shows the internal wiring of such an absorption meter.

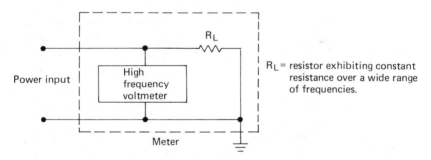

Figure 4.22 Internal wiring of an absorption meter.

Load resistors are available which exhibit constant resistance for frequency ranges from dc to 4.0 GHz. Although resistors with fairly constant frequency-resistance characteristics are available, this type of meter is usually limited to frequencies below 500 MHz. For higher frequencies well into the microwave region, other types of absorption meters are used.

The two other types of absorption meters which are extensively used in the band of frequencies from 500 MHz to 40 GHz are the calorimeter and bolometer power meters.

4.14 CALORIMETER POWER METER

The basic principle of operation of the calorimeter-type power meter is quite simple. A resistor is heated by the signal to be measured. This resistor is in a completely enclosed, well-insulated fluid bath. See Fig. 4.23. The insulation of the walls of the bath are such that very little heat is allowed to radiate from the bath. If the bath temperature is measured before and after the signal is applied, the difference in temperature is a measure of the heat developed by the resistor. Knowing the volume, the specific heat of the fluid, and the characteristics of the bath, the power can be calculated.

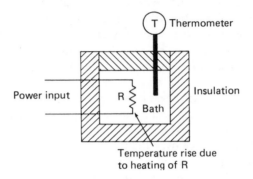

Figure 4.23 Basic calorimeter power meter.

The need to know the exact mass and specific heat of the calorimeter bath material, coupled with the extreme difficulty of obtaining complete thermal isolation from the environment, restricts the usage of the basic calorimeter power meter. Some of these difficulties are avoided in an instrument called the substitution calorimeter. In this instrument the signal is applied to the terminating resistors in the bath, and the equilibrium temperature of the bath is recorded. Now the signal is removed and dc or low-frequency ac is applied until the same equilibrium temperature is reached. The power can easily be measured with an accurate voltmeter. Figure 4.24 shows the basic substitution power meter.

Another calorimetric-type instrument is the comparison flow calorimeter. In this instrument the fluid is made to pass in separate streams by the input load and the comparison load. The power to the comparison load is a function of the difference in temperature between the loads since it is the amplified output from an unbalanced bridge. The meter measures the power to the comparison load which balances the bridge. In as much as the gauges are the same and the heat transfer characteristics are the same, the power in each load is the same; therefore the meter can be calibrated in input power. Figure 4.25 shows a simplified diagram of the measuring circuit of the

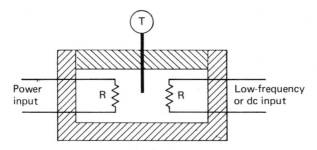

Figure 4.24   Basic substitution power meter.

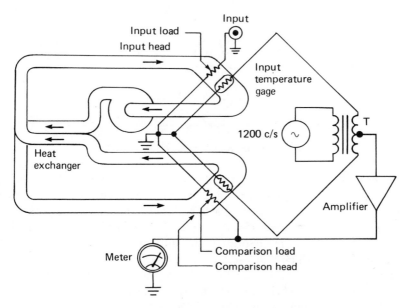

Figure 4.25   Simplified diagram of the Hewlett-Packard model 434A comparison flow calorimeter power meter.

Hewlett Packard Model 434A comparison flow calorimeter power meter. In the diagram we see the input load with its associated temperature gauge and the comparison load with its gauge. The two gauges constitute two arms of an ac bridge. The remaining bridge arms are really the center tap winding of the transformeter (T). When the bridge is unbalanced due to the difference in temperature between the two loads, there will be an error voltage between the input of the amplifier and ground. This voltage is amplified and fed back to the comparison load, where it is monitored by the meter. When balance is achieved the power in the comparison load must be equal to the power in the input load, so that the meter which is calibrated in power units reads

input power. One of the most frequently used absorption-type meters is
called the bolometer bridge.

4.15 BOLOMETER

The bolometer is basically a bridge circuit where one of the arms contains a
temperature-sensitive resistor. See Fig. 4.26. The temperature-sensitive resistor is placed in the field of a microwave signal whose power is to be measured.

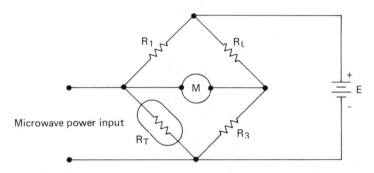

Figure 4.26 Basic bolometer bridge circuit.

The power is absorbed by the resistor, and the heat generated causes a change
in resistance. This change of resistance is measured with a bridge circuit.
Figure 4.27 shows the Hewlett Packard power meter and a simplified
schematic of the measuring circuit. Referring to the schematic of Fig.
4.27, we see that the combination of the differential amplifier and bridge
form an oscillator which will oscillate at a particular amplitude when the
bridge is unbalanced. The sensitive resistive element (R_T) absorbs the power
and heats itself until the bridge is very nearly balanced. The dc biased is then
adjusted until exact balance is achieved. Now the element is placed in the
microwave field. The element absorbs the power and is heated, which causes a
change in its resistance, thereby unbalancing the bridge. The imbalance is in a
direction opposite to that of the initial cold resistance. This action causes the
output from the oscillator to decrease so as to effect a balance. The EVM
circuit measures the amount of this power decrease from the oscillator and
displays the measurement as a power increase which is that supplied by the
microwave field.

Bolometer bridges use two types of elements: barretters and thermistors.
Barretters are metallic wires or film in which the temperature coefficient of
resistance is positive; that is, the higher the temperature, the higher the resistance. Thermistors are resistors made of semiconductor material which has a

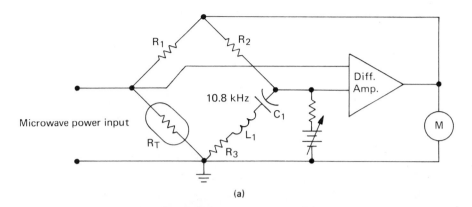

(b)

Figure 4.27 (a) Simplified diagram of the measuring circuit of a modern bolometer-type power meter. (b) Hewlett-Packard model 432 A power meter (courtesy of Hewlett-Packard, Palo Alto, Calif.).

negative temperature resistance coefficient; this means that the resistance of the element decreases as the temperature increases. In general barretters are very delicate, whereas thermistors are much more rugged. The bolometer is customarily used for power measurements in the range 0.01–10 mW. In the range of 10 mW to 10 W a comparison-type calorimeter such as the HP434A is quite suitable. Above 10 W, power measurements can be made by using attenuators and a low-power bolometer. Typical arrangements for measuring average power from 10 mHz to 40 GHz are shown in Fig. 4.28.

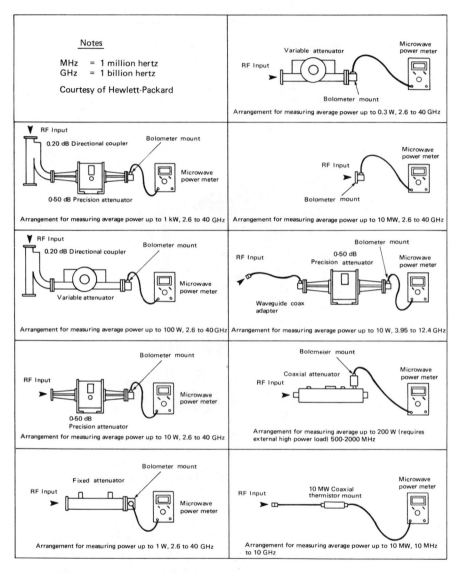

Figure 4.28 Typical arrangements for measuring average power from 10 mHz to 40 GHz.

4.16 SUMMARY

Power is defined as

$$\frac{\text{work}}{\text{time}} = \frac{W}{t} = P$$

Electrically, power is defined as $P = IE \cos \theta$, the unit being the watt. This is called average or true power. The angle θ is the phase angle between I and E, and $\cos \theta$ is called the power factor (F_p). Where $F_p = 1$ the circuit is purely resistive, and the total power delivered to the circuit is equal to the average power. Where there are many reactive elements $F_p < 1$, and the total power is not equal to the average power. There are three powers defined:

P_A = apparent power   (volt-amperes)

P_T = average or true power   (watts)

P_R = reactive power   (vars)

They are related by

$$P_A = \sqrt{P_T^2 + P_R^2}$$

where $P_R = P_{RL} - P_{RC}$, P_{RL} is the inductive reactive power, and P_{RC} is the capacitive reactive power.

Where the power factor is much less than 1, additional reactive elements must be added to correct the condition since the power companies will be delivering more power than they are being paid for.

Low-frequency wattmeters use electrodynamometer movements. The power in three-phase ac systems can be measured using either three wattmeters or two wattmeters in situations where the circuit junctions are not available. A device which uses a modified electrodynamometer movement is the power-factor meter, which can measure the power factor of either single- or three-phase circuits. The watt-hour meter is an instrument which measures total energy delivered to a system. The power companies use it to keep track of the energy delivered to a system and charge the customers for the power delivered. Measurements of power at higher frequencies, especially in the microwave range, require specialized instruments. In the range 0.01–10 MW, power is measured using a device called the bolometer bridge. Above that range, one must resort to using attenuators with the bolometer or a calorimeter-type instrument.

REVIEW QUESTIONS

1. Define power in terms of
   a. Mechanical units.

   b.   Voltage and resistance.

   c.   Voltage, current, and power factor.

2. Compute the power delivered to the circuits of Fig. 4.29.

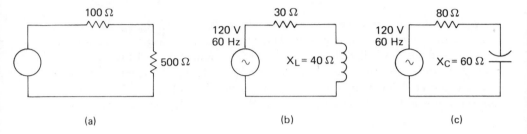

  (a)   (b)   (c)

Figure 4.29

3. For a given system, $E = 120$ V, $I = 5$ A, and $F_p = \cos\theta = 0.8$. Calculate the following powers: P_A, P_T, and P_R.

4. For a given system, $P_A = 4{,}000$ V·A and $P_R = 1{,}000$ var. calculate the average power P_T.

5. Show how you would connect a wattmeter in the circuit shown in Fig. 4.30 to measure the power dissipated in R_4. Indicate when the wattmeter would be connected to measure the total power delivered.

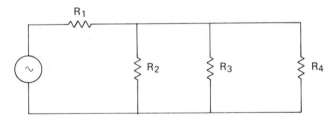

Figure 4.30

6. Explain the principle of operation of the power-factor meter.

7. The dials of a watt-hour meter are shown in Fig. 4.31. If electricity is sold at 2.5 cents/kWh, what would be the cost of the amount indicated?

8. Explain the difference between a barretter and a thermistor.

9. Explain the basic principle of operation of a bolometer.

Figure 4.31

5

OSCILLOSCOPE
MEASUREMENTS

5.1 INTRODUCTION

An oscilloscope (referred to as a CRO or scope) is a universal measuring instrument capable of measuring a wide variety of rapidly changing electrical signals. The signals may be repetitive or occur only once and last a fraction of a microsecond.

In the previous chapters the instruments (meters) discussed are accurate and reliable only for measuring dc or ac signals. One must realize that many signals occur in the world of electronics which are not dc or ac, such as the incoming TV signal (Fig. 5.1) into a television receiver. The universal instrument which is capable of measuring these electronic signals and presenting a picture of the exact signal as a function of time is the oscilloscope. The oscilloscope graphs the changes in signal amplitude on its vertical axis and how long the event lasts on its horizontal axis. With the oscilloscope the user can determine the signal's polarity (positive or negative), amplitude, and duration. Depending on scope sophistication, there are general-purpose scopes capable of viewing one signal only that operate from 100 Hz to 500 kHz, while others operate from dc to 100 MHz and are capable of viewing four different signals simultaneously. Oscilloscopes are also available (called storage scopes) which can store the signal trace until observed or until a second signal can be stored for direct comparison.

This chapter basically covers two areas: basic oscilloscope operation

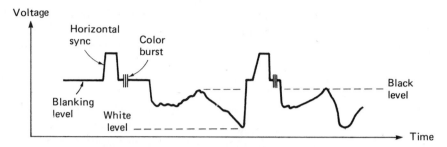

Figure 5.1   Part of television signal waveform.

and measurement techniques utilizing the oscilloscope. In discussing basic oscilloscope operation stress will be placed on those scope features which the student may have difficulty in understanding, such as sweep synchronization, the difference between the alternate and chop modes, sweep magnification, and the delay sweep feature.

Regarding measurement techniques, we shall initially discuss basic measurements of voltage, frequency, and phase and shall proceed to sophisticated measurements such as square-wave and pulse testing and alignment of an IF circuit.

5.2   BASIC OSCILLOSCOPE OPERATION

The heart of the oscilloscope is the cathode ray tube (CRT), which produces the electron beam and focuses and accelerates it toward a phosphor screen. Upon striking the screen, the electron beam converts its kinetic energy to light energy. Refer to Fig. 5.2. With zero voltage applied to the deflection

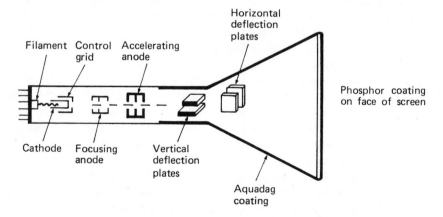

Figure 5.2   Internal construction of a cathode ray tube.

plates, the electron beam will strike the center of the CRT, causing a small luminous spot. With a voltage applied to the deflection plates (horizontal-vertical), the electron beam will be deflected horizontally and vertically, directly proportional to the voltage at the specific plates. The beam may now be positioned horizontally or vertically by applying a dc voltage to the deflection plates.

Since the beam movement is directly proportional to the voltage at the deflection plates, let us examine what happens if a sweep (ramp or sawtooth) voltage, as shown in Fig. 5.3, is applied across the horizontal plates. The

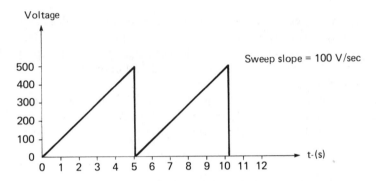

Figure 5.3   Typical ramp, sawtooth, sweep waveform.

beam is initially positioned at the extreme left of the CRT face, and as the ramp voltage increases the spot will be deflected toward the right. If the CRT has a 5-in. diameter and requires 100 V across its plates for a deflection of 1 in., the beam will deflect 5 in. toward the right with the application of a 500-V ramp signal. With the ramp slope increasing at a linear rate of 100 V/s, the spot will move from left to right at a rate of 1 in./s, or it will take 5 s to sweep the beam across the CRT face. Refer to Fig. 5.4. At the completion of 5 s, the ramp voltage suddenly falls to 0 V, causing the beam to retrace back to its initial position.

If a 500-V sinusoid (period = 5 s) is now applied to the vertical plates (Fig. 5.5) simultaneously with the ramp voltage at the horizontal plates, the beam will trace on the screen an exact presentation of the sinusoidal voltage. If the sinusoidal period is reduced to 2.5 s (frequency doubled), a sweep voltage of 100 V/s will cause two cycles of the sinusoid to appear on the screen. With frequency equal to the inverse of the period, it can be restated that an input signal frequency twice the sawtooth frequency will cause two cycles to appear on the screen. By regulating the ratio of sweep frequency to input signal frequency, portions of the input signal or many cycles may be viewed on the screen.

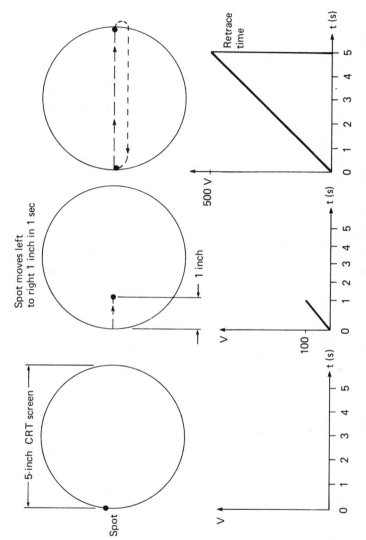

Figure 5.4 Spot deflection for a ramp voltage applied to horizontal plates with zero voltage on the vertical plates.

105

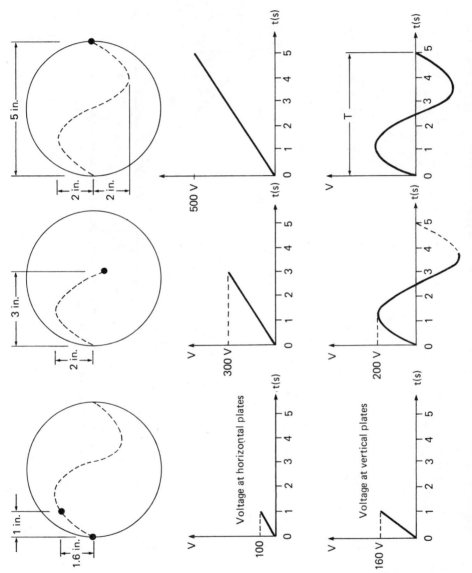

Figure 5.5 Oscilloscope beam movement as a function of voltages on horizontal and vertical plates.

106

5.3 SIMPLIFIED FUNCTIONAL BLOCK OF AN OSCILLOSCOPE

A simplified block diagram for a basic oscilloscope is shown in Fig. 5.6. The waveform to be observed is fed into the vertical input, where it is amplified and fed to the vertical deflection plates of the cathode ray tube. Simultaneously the sweep generator is triggered-on via the sync circuit, producing a linear sweep voltage. The sweep voltage is amplified and applied to the horizontal deflection plates. A stable display of the waveform will now appear on the CRT screen. Each of the blocks will now be discussed in detail.

5.4 VERTICAL AMPLIFIER ATTENUATOR

For the scope to be capable of measuring voltages varying from millvolts to hundreds of volts, the signal to be measured is initially fed into a calibrated step attenuator. The attenuator is calibrated in volts per division, with a typical attenuation factor of 500 to 1, in 9 to 10 ranges. The attenuator output is followed by a voltage amplifier with a gain of approximately 2000:1. The attenuator-amplifier combination results in a scope with a vertical deflection varying typically from 10 mV to 20 V/division. For example, if a 100-mV signal is to be viewed, the calibrated attenuator is set to 20 mV/division, resulting in a vertical deflection (of the signal) on the CRT trace of 5 divisions.

The frequency response of the vertical amplifier often determines the general class of oscilloscope. Available are two general classes: low frequency and high frequency. The low-frequency scope has a vertical amplifier with a frequency response bandwidth of dc to 10 MHz, while a high-frequency scope may go as high as 350 MHz. With high-frequency scopes often intended for pulse measurements, the rise time (t_r) capability is often specified. A relationship between bandwidth and rise time exists:

$$\text{bandwidth (BW)} = \frac{0.35}{t_r}$$

where bandwidth is in megahertz and t_r is in *microseconds*. A high-frequency scope with a rise time of 1 ns has a frequency response of 0.35/1 ns = 350 MHz. Refer to Section 5.19 for a further explanation of pulse characteristics.

5.5 SWEEP SYNCHRONIZATION CIRCUIT

To present a stationary pattern on the CRT screen the sweep voltage frequency must be equal to a fixed multiple of the vertical input signal being viewed. This is accomplished by the synchronization (trigger generator) circuit, which initiates the sweep.

107

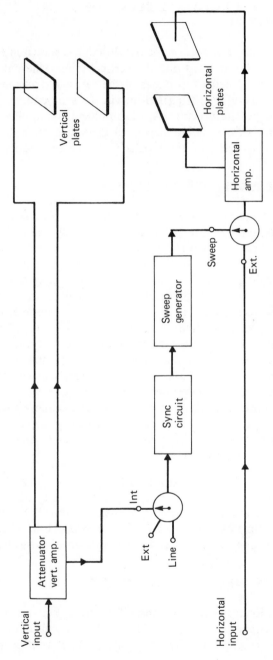

Figure 5.6 Simplified block diagram of an oscilloscope.

Trigger generators incorporate several trigger sources, plus a variable comparator for setting the desired trigger level. Typical trigger sources are labeled "internal," "external," and "line." With the trigger mode switch set to "internal," the trigger generator signal (Fig. 5.7) is obtained from the

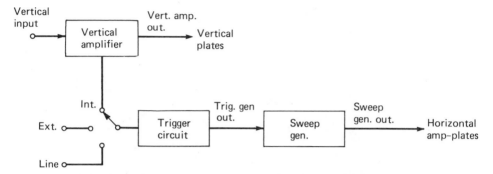

Figure 5.7   Simplified block diagram of synchronization-sweep circuit.

vertical signal being measured. This synchronization mode is used in conjunction with the trigger level slope controls, and by proper adjustment the trigger signal can be set to originate when the input signal is going positive, negative, or at any particular voltage level. With the trigger level control set to "automatic," the trigger generator will automatically provide a trigger signal to the sweep generator with no vertical input signal. This function thus gives a base-line presentation without flicker and allows for easy base-line setting. With the application of an input signal, the trigger generator pulse output occurs when the input signal is 0 V. Figures 5.8(a) and (b) demonstrate the "auto" trigger level of operation for two different sweep frequencies.

For a trigger signal to occur when the input voltage is at a particular voltage level other than zero, the trigger level is moved to + or − position. Refer to Fig. 5.9 for typical CRO output waveforms for different trigger levels and mode settings.

The sweep oscillator may also be synchronized to an external signal (trigger mode on "ext.") or synchronized to the 60-Hz line frequency which results with the trigger mode switch set to "line."

5.6   SWEEP GENERATOR–HORIZONTAL AMPLIFIER

To deflect the electron beam horizontally across the CRT face a linear sweep voltage is required. The sweep generator produces this voltage and, depending on the type of scope, may feature a calibrated sweep time of from 10 ns to 5 s/division, with a time accuracy range of better than 3%. Depending on

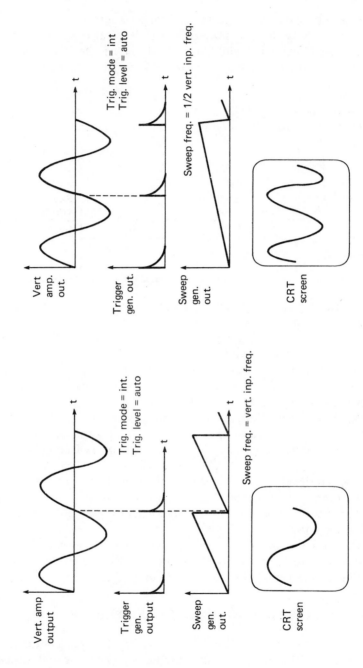

Figure 5.8 Synch-sweep waveforms.

110

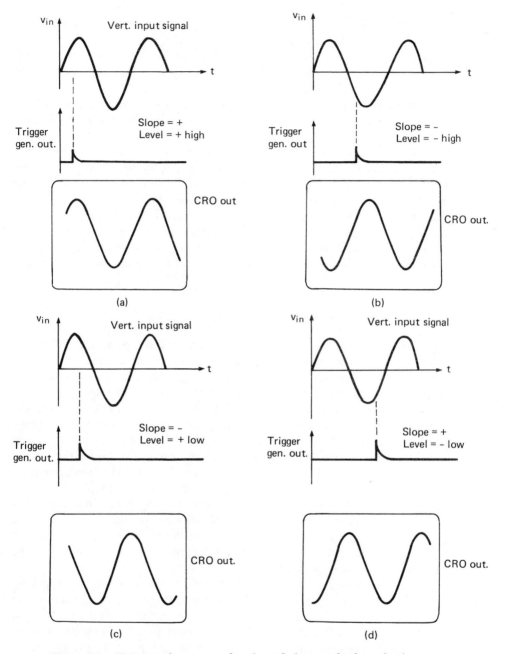

Figure 5.9 CRO waveforms as a function of slope and trigger level settings.

the trigger mode-level setting, the sweep generator may be triggered by the voltage being displayed or by an external signal.

During the sweep time, a gate voltage is applied to the CRT control grid (Z axis), turning on the beam. During retrace time, the control grid is gated off, blanking out the trace.

The sweep generator output voltage is unable to directly drive the horizontal plates but must be initially amplified. This is accomplished by feeding the sweep output via a sweep selector switch to the horizontal amplifier. With the sweep selector control in its "on" position the sweep generator is connected directly to the horizontal amplifier, producing a sweep across the CRT screen. With the sweep selector switch set to external (or "off"), the horizontal amplifier input is connected instead to the "horizontal input" jack. In this position the beam will remain stationary. By feeding in an external signal into the horizontal amplifier simultaneously as another signal is fed into the vertical input, the scope will act as an X-Y recorder. This mode of operation may be used for measuring phase and frequency and for aligning an IF amplifier.

5.7   DUAL TRACE SCOPE FEATURE

Many oscilloscopes have the feature of dual trace displays, which enables the operator to perform time and amplitude comparisons between two waveforms. This is accomplished with one sweep generator and two vertical amplifiers operating in the *alternate* or *chop* mode. The alternate mode displays one vertical channel for a full sweep and the other channel on the next sweep. The result is that the output of each vertical channel is alternately displayed. This mode of operation (alternate) is commonly used for viewing high-frequency signals, where sweep speeds are much faster than CRT phosphor decay time. When low-frequency signals are to be viewed, requiring a slow sweep, the alternate mode of operation is objectionable due to the extreme flicker. This is overcome by utilizing the chop mode. In this mode, a small time segment of the sweep will be allotted to one vertical channel, with the next time segment allotted to the second channel. The result is that each vertical channel is composed of small chopped segments, which merge to appear continuous to the eye.

Both the alternate and chopped modes of operation are obtained by utilizing an electronic switch with two vertical preamplifiers (containing attenuators and positioning controls) and one vertical amplifier. Refer to Fig. 5.10. In the alternate mode, the electronic switch alternates between preamplifiers every sweep period, while in the chop mode, the electronic switch oscillates between channels at a fixed rate of approximately 100 kHz. Refer to Figs. 5.11(a) and (b) for sweep and gate voltages for alternate and chop modes of operation.

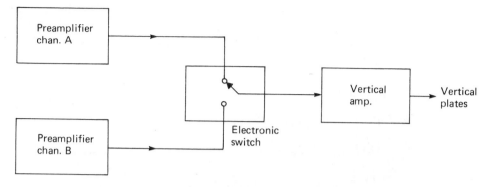

Figure 5.10 Block diagram for generating a dual trace utilizing an electronic switch. *Alternate mode:* electronic switch will alternate between each amplifier for a sweep period. *Chop mode:* electronic switch will alternate between each amplifier for a fixed interval of approximately 5–10 μs independent of the sweep period.

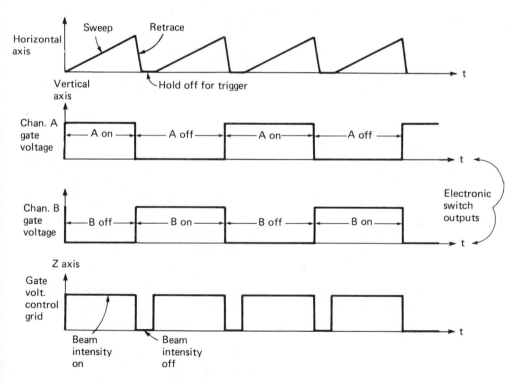

Figure 5.11 (a) Timing relationship for a dual channel vertical amplifier in alternate mode operation.

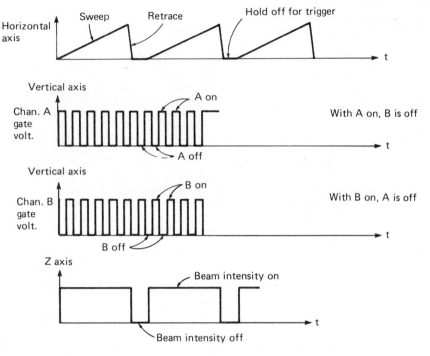

Figure 5.11 (b) Timing relationship for a dual channel vert. amp. in chop mode operation.

5.8 SWEEP MAGNIFICATION FEATURE

By increasing the horizontal amplifier's gain, it is possible to expand the screen display for a detailed observation of the waveform. This is accomplished by the sweep magnification control which may expand the sweep up to 10 times. To determine time measurements using the sweep magnifier, the horizontal sweep in seconds per division is converted to

$$\frac{\text{seconds per division}}{\text{Magnification factor}}$$

For example, with the sweep set to 10 μs/division, the center pulse waveform of Fig. 5.12(a) is

$$(10 \ \mu \text{ s/div.})(3 \text{ div.}) = 30 \ \mu s$$

With the sweep magnifier set to X 10 [Fig. 5.12(b)], the pulse width is

$$\frac{(10 \ \mu \text{s/div.})(30 \text{ div.})}{10} = 30 \ \mu s$$

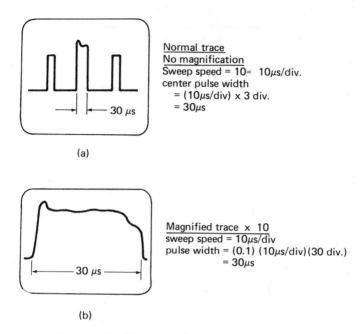

(a)

Normal trace
No magnification
Sweep speed = 10= 10μs/div.
center pulse width
= (10μs/div) x 3 div.
= 30μs

(b)

Magnified trace x 10
sweep speed = 10μs/div
pulse width = (0.1) (10μs/div)(30 div.)
= 30μs

Figure 5.12   Waveform display of a magnified trace.

5.9   DELAYED SWEEP FEATURE

A measurement which occurs on occasion is the determination of the rise time of a signal which doesn't originate at the beginning of a cycle. For example, the rise time of pulse 2 of a pulse pair depicted in Fig. 5.13(a) is to be measured. With the sweep speed set for measurement of the rise time t_r (sweep speed, 0.1 μs/division), the waveform of pulse 1 appears on the screen, while pulse 2 [Fig. 5.13(b)] will be off the screen. This measurement may be performed with a scope with a delayed sweep feature.

The principle of operation is as follows: The CRO contains two linear calibrated sweeps, a main sweep and a second sweep referred to as the delayed sweep. The main sweep is initiated by its trigger pulse at time t_0 [Fig. 5.14(a)] and will continue until reaching a comparator level set by the delay multiplier ("delay div.") control. At this time t_1, the delay generator will be triggered on (delay generator trigger set to "auto"), intensifying a portion of the original display. If the scope sweep control is now set to its delay position, the original display will be replaced with a display of the intensified portion expanded on the screen. Refer to Fig. 5.14(b). By adjustment of the delayed sweep speed, any portion of the waveform can be ex-

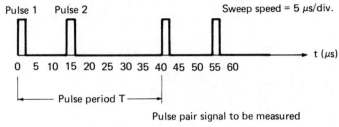

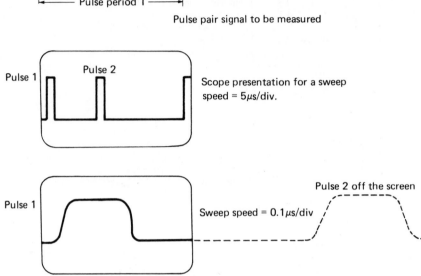

Figure 5.13 Pulse pair display for two different sweep speeds.

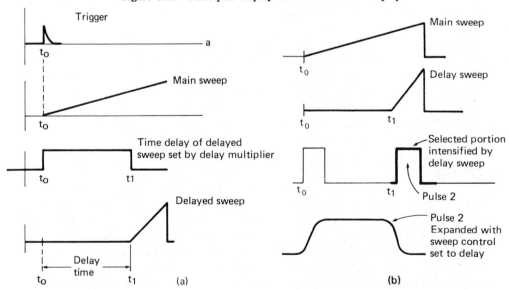

Figure 5.14 (a) Sweep waveforms of a CRO with a delay feature. (b) Sweep expansion of a waveform by utilizing delay feature.

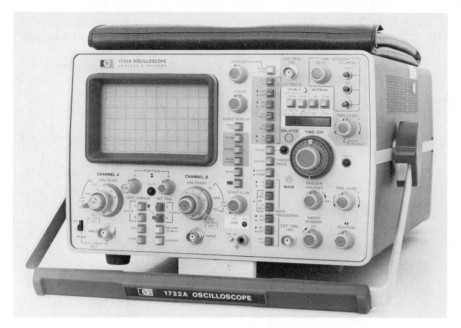

Figure 5.15   Dual trace oscilloscope. Courtesy of Hewlett-Packard.

panded for proper measurements. Refer to Fig. 5.15 for a photo of a dual trace oscilloscope with delay features.

5.10   SAMPLING OSCILLOSCOPE

The sampling scope permits measuring of signals in the gigahertz (10^9 Hz) range. The scope relies upon a technique very similar to that of a stroboscopic light in providing visual observation of a rapid motion.

The waveform is not continuously measured, but discrete samples of signal amplitude at related intervals are measured and synthetically reproduced into a complete signal. The signal must be repetitive with thousands of samples required in order for the signal to appear continuous on the CRT trace. Refer to Fig. 5.16 for a repetitive waveform reconstructed via the sampling technique.

5.11   OSCILLOSCOPE PROBES

5.11.1   Compensated Attenuator Probe

The input impedance of a general-purpose oscilloscope is 1 MΩ in shunt with a 10- to 80-pF capacitance. A wire connected from the scope vertical input to the circuit under test is adequate for low-frequency measurements but suffers from stray signal and 60-Hz pickup. If a coaxial cable is used to shield the probe from stray signals, it will significantly increase the input

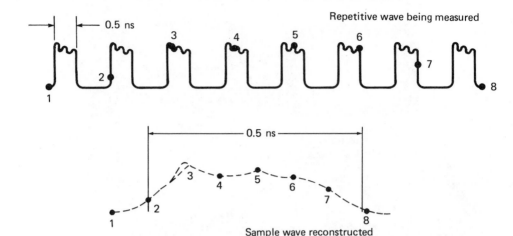

Figure 5.16   Reconstruction of a signal by sampling technique.

shunt capacitance, loading down the circuit. To overcome this a compensated (attenuator) probe may be used. Typical probes available have 10 to 1, 50 to 1, and 100 to 1 attenuation ratios.

The compensated probe presents a higher input resistance and lower shunt capacitance, attenuating all frequencies equally. A circuit of a typical 10:1 compensated probe, as shown in Fig. 5.17(a), increases the input resistance to 10 MΩ with a shunt capacitance of 10 pF. A photograph of a typical probe is shown in Fig. 5.17(b). Refer to Section 5.12 to determine if the probe is properly compensated.

5.11.2   Demodulator Probe

The demodulator probe is used for converting an RF signal into a dc voltage, with its output voltage directly proportional to the peak RF signal. Refer to Fig. 5.18 for a typical probe circuit. This type of probe is primarily used for signal tracing and for alignment of receivers. Refer to Section 5.20, where this probe is used with a swept oscillator for quick alignment of an IF amplifier.

5.11.3   Active Probe

The active probe consisting of an FET input is primarily used for high-frequency measurements (very small rise time) of extremely low-level pulse signals. An active probe will have a lower shunt capacitance than a compensated attenuator, with the added feature of zero signal attenuation. The disadvantages are cost and a limited dynamic signal capability. The P620 (XI) active probe (manufactured by Tektronix) has a shunt capacity of 0.4 pF and is capable of measuring a rise time t_r of 0.4 ns. The maximum signal

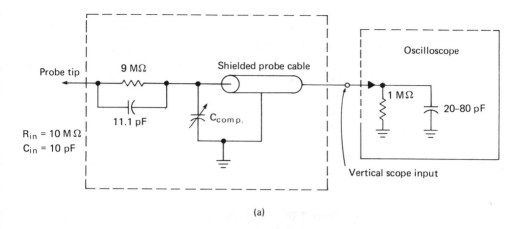

(a)

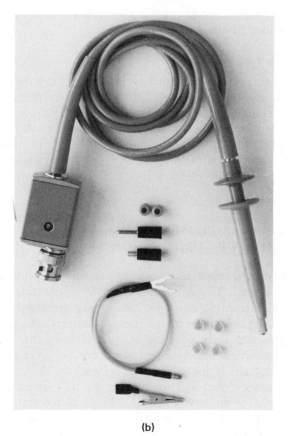

(b)

Figure 5.17 (a) Compensated 10:1 probe. (b) Compensated probe (courtesy of Hewlett-Packard).

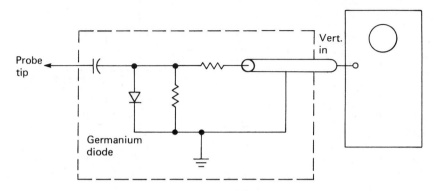

Figure 5.18   Typical demodulator probe.

voltage it can effectively measure is pulses with a peak amplitude of ± 0.6 V, with a maximum dc offset voltage of 5.6 V.

5.11.4 Current Probe

By utilizing a current probe (with associated amplifier), one may use a scope for current measurements. A useful probe for current measurements without disturbing the circuit is the split-core unit. The core slides back allowing the current-carrying lead to be inserted without breaking the circuit. A typical current probe with its associated amplifier can measure currents from 1 mA to 1 A with a frequency range from dc to 50 MHz. Measuring current by indirectly measuring the voltage across a fixed resistor is discussed in Section 5.16.

5.12   VERTICAL AMPLIFIER CALIBRATION AND ATTENUATOR COMPENSATION

Prior to utilizing the scope in performing measurements the scope and probe are to be calibrated. This is accomplished by connecting the scope via its probe to the internal square-wave calibrating voltage of the scope. The vertical sensitivity, sweep speed, and trigger controls are adjusted to produce a stationary square wave covering at least 60% of the scope face. *Note*: Make sure the vertical amplifier and sweep control are set to calibrate. The square-wave voltage is measured [Fig. 5.19(a)] by determining its vertical height in divisions (div.) and multiplying by its vertical amplifier setting, which is in volts (or millivolts) per div. The measured voltage is equal to the volts per div. of amplifier $\times$ square-wave height (in div.) $\times$ attenuation of probe (if any). For example, if a 10:1 probe is used and the vertical amplifier is set to 0.05 V/div., the square-wave voltage is equal to 0.05 V/div. $\times$ 2 div. $\times$ 10 = 1 V p to p. The measured output (square-wave) voltage should

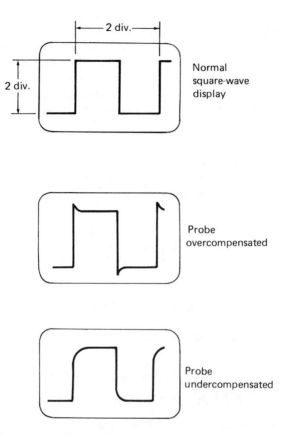

Figure 5.19 Presentation of internal square wave for scope and probe calibration.

be equal to the calibrated output, and any discrepancy is to be corrected by adjusting the gain of the vertical amplifier. Refer to a scope manual for proper adjustment. To minimize circuit loading it is good practice to use an attenuator probe when possible. Many times the probe, which is a compensated attenuator, requires adjustment. If the calibrating square wave has rounded corners or an overshoot [Figs. 5.19(b) and (c)], the probe requires adjustment. Refer to the manual on how to compensate the probe. The majority of measurements which follow will be made utilizing a 10:1 probe.

5.13 VOLTAGE MEASUREMENTS

When voltage measurements are to be made, the best results are obtained by measuring the peak-to-peak voltage (this measurement is also the simplest to perform). The peak voltage and average or rms values are obtained by

simple conversions. As described in Section 5.12, the waveform to be measured is fed into the vertical input via a 10:1 probe. The vertical amplifier
sensitivity, sweep speed, and triggering controls are adjusted for a stationary
waveform covering as many vertical divisions as possible. Adjust the vertical
and horizontal position controls such that the peak of the signal intersects
the scope graticule horizontally and vertically. The peak-to-peak voltage
(Fig. 5.20) is the vertical amplifier sensitivity in volts/div. × the number of

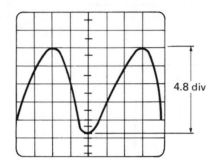

Figure 5.20   Scope sensitivity = 0.01 V/Div. 10:1 probe used.

vertical divisions × the probe attenuation. For the sinusoidal signal shown
in Fig. 5.20,

$$e_{p\,to\,p} = \text{No. of Div.} \times \text{vertical sensitivity (volts/Div.)} \times 10$$

$$= (4.8)(0.01)(10) = 0.48\ v_{p\,to\,p}$$

$$e_{peak} = \frac{1}{2}\,e_{p\,to\,p} = \frac{0.48}{2} = 0.24\ \text{V peak}$$

$$e_{rms} = \frac{e_{p\,to\,p}}{2.828} = \frac{0.48}{2.828} = 0.1697\ \text{V rms}$$

5.13.1   Direct Current Voltage Measurements

With a 10:1 probe, connect the dc voltage to be measured into the CRO
vertical input. Set the voltage mode switch to "gnd" (or "ac" if no "gnd"
is available), and adjust the vertical position control for a centered trace.
Move the mode switch to dc, and note the trace deflection. If the trace deflects
upward, the dc voltage is positive, and if the trace moves downward, it is
negative. Reset the mode switch to gnd, and position the trace to the lowest
or uppermost screen graticule position depending on the dc voltage polarity.
Reset the mode switch to dc, and adjust the vertical sensitivity for a maximum
screen deflection.

$$V_{dc} = \text{trace deflection (div.)} \times \text{vertical sensitivity (volt/div.)} \times 10$$

5.13.2 Pulsating dc Voltage Measurements

A frequent measurement to be made is the measurement of an ac riding on a dc level. A typical application is determining the output voltage of a power supply (Fig. 5.21) consisting of an ac ripple voltage riding above a dc voltage.

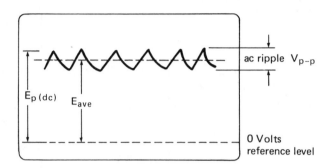

Figure 5.21

To perform this measurement, set the voltage mode to ac input, and with a 10:1 probe, measure the peak-to-peak ac ripple voltage. The vertical sensitivity should be set for the ac ripple voltage to occupy a majority of the screen. Set the voltage mode switch to gnd, and adjust the horizontal trace to the lowest graticule position. This point is now our ground (zero) reference position. Set the mode switch to dc, and readjust the vertical sensitivity in order to determine the peak voltage excursion $[E_{p(dc)}]$ above the reference gnd. The average value is the peak voltage excursion less half the peak-to-peak ac voltage.

Determining the rms and average value of various waveforms Many times the rms or average value of various voltage waveforms are to be determined when the peak voltage is known. Table 5.1 lists the relationships among the peak, rms, and average values for different waveforms.

5.14 FREQUENCY MEASUREMENTS

5.14.1 Sweep Method for Measuring Frequency

A signal frequency is determined by initially measuring its period T and calculating the frequency by

$$f(\text{hertz}) = \frac{1}{T} \text{ (seconds)}$$

The procedure is as follows: The waveform to be measured is fed into the vertical input, and the scope vertical sens., sweep speed, and triggering con-

Table 5.1

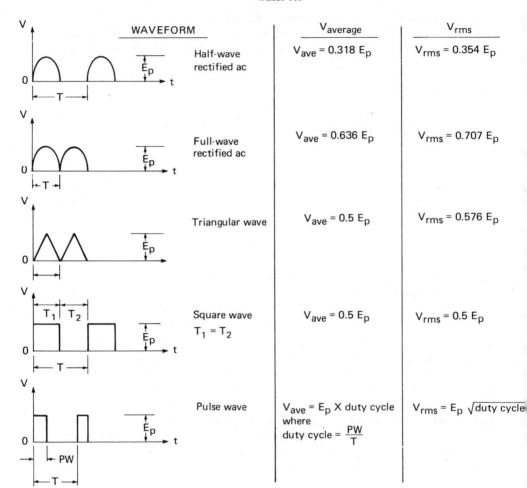

WAVEFORM		$V_{average}$	V_{rms}
Half-wave rectified ac	E_p	$V_{ave} = 0.318\ E_p$	$V_{rms} = 0.354\ E_p$
Full-wave rectified ac	E_p	$V_{ave} = 0.636\ E_p$	$V_{rms} = 0.707\ E_p$
Triangular wave	E_p	$V_{ave} = 0.5\ E_p$	$V_{rms} = 0.576\ E_p$
Square wave $T_1 = T_2$	E_p	$V_{ave} = 0.5\ E_p$	$V_{rms} = 0.5\ E_p$
Pulse wave	E_p	$V_{ave} = E_p \times$ duty cycle where duty cycle $= \dfrac{PW}{T}$	$V_{rms} = E_p \sqrt{\text{duty cycle}}$

trols are adjusted for a stable waveform of at least one complete cycle covering a majority of the CRO face. With the number of horizontal divisions for a complete cycle determined, the period is found by

period T (s) = (number of horizontal div. for 1 cycle) $\times$ [sweep speed (s/div.)]

Referring to Fig. 5.22, we see that a complete period is found to be 6 div. with the sweep speed set to 10 μs/div.:

$$\text{period } T = 6 \text{ div.} \times 10\ \mu\text{s/div.} = 60\ \mu\text{s}$$

$$\text{frequency} = \frac{1}{T} = \frac{1}{60 \times 10^{-6}} = \frac{10^6}{60} = 16{,}667\ \text{Hz}$$

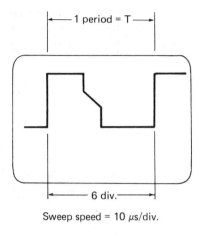

Sweep speed = 10 μs/div.

Figure 5.22

The above method of measuring frequency is accurate to approximately ±3%, this being the sweep generator accuracy.

5.14.2   Comparator (Lissajous) Method for Measuring Frequency

A more accurate method for measuring frequency is to utilize the scope as a frequency comparator. This may be accomplished by feeding in the unknown frequency into the vertical input, with a known frequency fed into the horizontal input. With the sweep generator set to off, and if both inputs are at the same frequency, a circular (Lissajous) pattern (Fig. 5.23) will appear on the screen. This method of measuring frequency is as accurate as the known frequency. If the horizontal frequency input is obtained from a frequency synthesizer which has an accuracy of 0.001% with a frequency

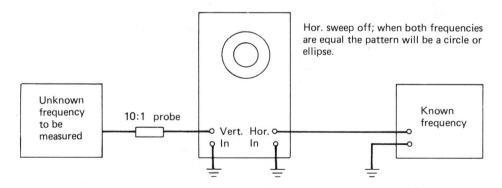

Figure 5.23   Comparison (Lissajous) method for measuring frequency.

variability of from 0.01 Hz to 100 MHz in 0.01-Hz steps, the scope can be utilized to measure frequency with the accuracy of the frequency synthesizer. If the frequency to be measured is higher than the known frequency range, Lissajous patterns as shown in Fig. 5.24 will result. The frequency ratio is found by the expression

$$f_v = \frac{t_h f_h}{t_v}$$

where f_v = frequency into vertical input,

   f_h = frequency into horizontal input,

   t_h = number of loops which touch the horizontal tangent line,

   t_v = number of loops which touch the vertical tangent line.

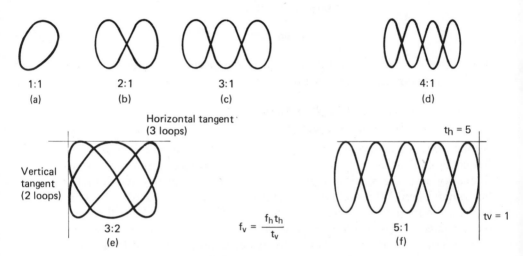

Figure 5.24   Various frequency patterns available when f_v is larger than f_h.

Referring to Fig. 5.24(f), with $t_h = 5$ and $t_v = 1$, the unknown frequency can be five times greater than the known frequency and still be measured.

5.15   PHASE MEASUREMENTS

5.15.1   *Alternate Sweep Method for Measuring Phase*

The phase relationship between two sinusoids may be directly measured by viewing both waveforms on the oscilloscope and determining the delay time T_d between the two waveforms. The scope is set to alternate mode (for viewing two waveforms simultaneously), with each vertical amplifier sensitivity and trigger control adjusted for two stationary sinusoids. Refer to

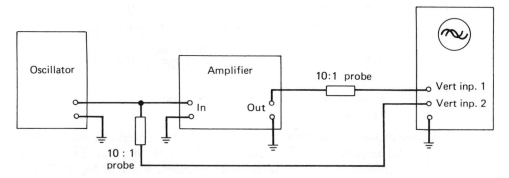

Figure 5.25 Test setup for measuring the phase shift of an amplifier utilizing the alternate sweep method.

Fig. 5.25 for the test setup for measuring the phase shift introduced by an amplifier. The sweep speed is initially adjusted such that the period T of the sine wave is measured. The sweep speed is then increased for accurately determining the delay time T_d between the two sinusoids. Refer to Figs. 5.26(a) and (b).

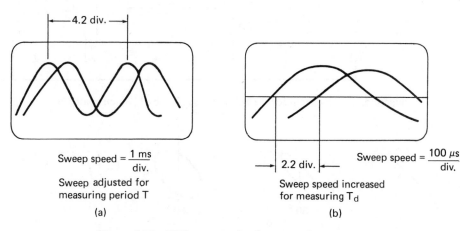

Figure 5.26 CRO presentation for measuring phase.

The phase is determined by

$$\text{phase delay (deg)} = \frac{T_d}{T} \times 360$$

Referring to Fig. 5.26(a), we find the period T by

$$T = (4.2 \text{ div.})(1 \text{ ms/div.}) = 4.2 \times 10^{-3} \text{ s}$$

The delay between the two sine waves [Fig. 5.26(b)] is found as

$$T_d = (2.2 \text{ div.})(100 \ \mu\text{s/div.}) = 220 \times 10^{-6} \text{ s}$$

$$\text{phase delay (deg)} = \frac{T_d}{T} \times 360 = \frac{220 \times 10^{-6}}{4.2 \times 10^{-3}} (360) = 18.86°$$

5.15.2   Phase Measurements Utilizing Lissajous Patterns

A second method for determining the phase shift is to feed one sinusoid into the vertical amplifier, with the other connected to the horizontal input. The scope sweep generator is set to the "off" position. Refer to Fig. 5.27. The vertical and horizontal gains are adjusted for a reasonably sized pattern on the CRT. Figure 5.28 shows a number of phase-shift patterns that are obtain-

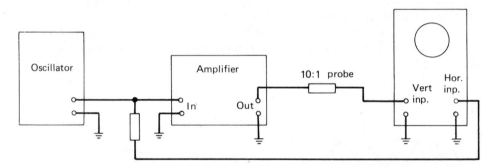

Figure 5.27   Test setup for measuring phase shift by Lissajous patterns.

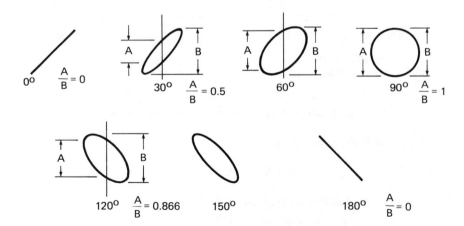

Figure 5.28   Typical Lissajous phase-shift patterns.

able. The phase shift (θ) in degrees is calculated by the following formula:

$$\theta = \sin^{-1}\left(\frac{A}{B}\right)$$

5.16 CURRENT MEASUREMENTS

The CRO cannot measure current directly, it having a high input impedance. An indirect method is to connect a noninductive resistor (carbon) whose value is known in series with the circuit whose current is to be measured. The Ohmic value of the sense resistor should be such a value that it doesn't alter the actual current flowing through the circuit. The voltage drop across the resistor is measured by an oscilloscope, and the current is found by using Ohm's law: $I = E/R$. A second method using a special current probe is described in Section 5.11.4.

5.17 POWER MEASUREMENTS

The oscilloscope can be utilized in measuring power by terminating the circuit whose power is to be measured with its proper load resistor (noninductive) and measuring the voltage across the load. The power output is found by

$$P_o = \frac{(E_{o(p \text{ to } p)}/2.828)^2}{R_L}$$

For example, for an RF signal generator, which is terminated with its load resistor of 50 Ω, an output voltage of 0.2 V peak to peak was measured. The generator power output is calculated as

$$P_o = \frac{(e_{o(p \text{ to } p)}/2.828)^2}{R_L} \frac{(0.2/2.828)^2}{50} = 0.1 \text{ mW}$$

5.18 SQUARE-WAVE TESTING

An amplifier may be tested by applying an ac signal of varying amplitude-frequency to determine its phase, gain frequency response, etc. This method of testing is long, tedious, and ineffective in determining the transient characteristics of the amplifier. An effective dynamic measurement which determines the amplifier linearity and transient and overall frequency characteristics is square-wave testing. This method of testing became widely used in the testing of low-medium-frequency amplifiers which were specifically designed to amplify signals making abrupt transitions.

It can be shown mathematically (Fourier analysis) that a square wave [pulse repetition frequency (PRF) = f_s] consists of a sinusoidal signal with an f_s (fundamental) with an infinite amount of odd harmonics. With the square wave having an amplitude E_p with a PRF f_s, it can be expanded as

Fourier analysis of a square wave =

$$\frac{E_p}{\pi}\left\{ \underset{\text{fundamental}}{\underbrace{\sin 2\pi f_s t}} + \frac{1}{3} \underset{\text{third harmonic}}{\underbrace{\sin [(3)(2\pi f_s t)]}} + \frac{1}{5} \underset{\text{fifth harmonic}}{\underbrace{\sin [(5)(2\pi f_s t)]}} + \cdots \right\}$$

Refer to Fig. 5.29, where a square wave is recomposed by summing the fundamental with the third, fifth, seventh, and ninth harmonics. The application of a square wave to an amplifier may therefore be considered as the equivalent of applying simultaneously an infinite number of sine waves. If the amplifier output response is a perfect square wave, the amplifier is linear, with a uniform frequency response reproducing all the individual harmonic components in the same relation in which they were present in the original wave. By the same token, the extent to which the square-wave output departs from its input will signify the amplifier's degradation with respect to linearity and frequency response.

The frequency response of an RC amplifier may be approximated by measuring the rise time and tilt of the output square wave. Refer to Fig. 5.30. The rise time of the square wave is approximately related to the amplifier high-frequency response by

$$f_{\text{high}}* = \frac{0.35}{t_r}$$

where f_h is in megahertz and t_r is in microseconds. Similarly, the tilt determines the low-frequency response and is determined by

$$f_L = \frac{Pf_s}{100\pi}$$

where $P = (E_d/E_p) \times 100$, called tilt or droop, in percent,

   f_s = pulse repetition frequency of square wave.

Depending on how exacting an amplifier's frequency response requirements are, a square wave may be approximated as consisting of its fundamental, third, fifth, seventh, and ninth harmonics. If the amplifier being tested has a specified frequency response from 20 Hz to 20 KHz, the minimum PRF of the square wave must be at least 20 Hz, with a PRF of 2.2kHz (20 kHz/ninth harmonic) for testing the high end of the amplifier.

5.18.1   Determining the Response of an Amplifier
by Square-wave Testing

Figure 5.31 depicts the test setup for performing a square-wave test on an amplifier. Adjust the scope controls in order to monitor both the input and

---

*f_{high} = bandwidth.

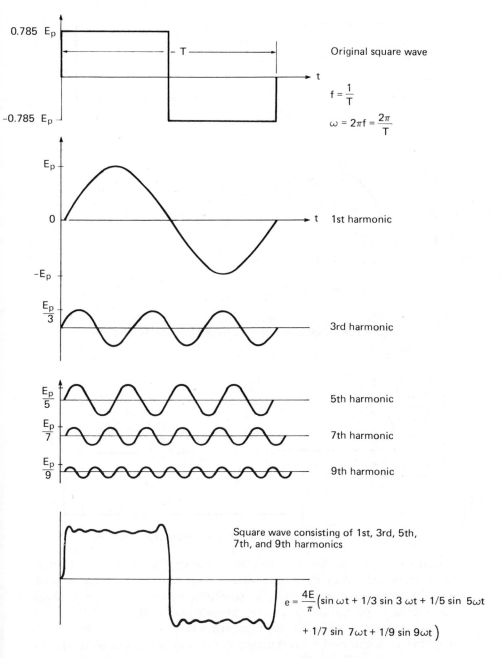

0.785 E_p

-0.785 E_p

Original square wave

$$f = \frac{1}{T}$$

$$\omega = 2\pi f = \frac{2\pi}{T}$$

E_p

0

$-E_p$

t   1st harmonic

$\dfrac{E_p}{3}$ 3rd harmonic

$\dfrac{E_p}{5}$ 5th harmonic

$\dfrac{E_p}{7}$ 7th harmonic

$\dfrac{E_p}{9}$ 9th harmonic

Square wave consisting of 1st, 3rd, 5th, 7th, and 9th harmonics

$$e = \frac{4E}{\pi}\left(\sin \omega t + 1/3 \sin 3\,\omega t + 1/5 \sin 5\omega t + 1/7 \sin 7\omega t + 1/9 \sin 9\omega t\right)$$

Figure 5.29 Composition of a square wave from its harmonic components.

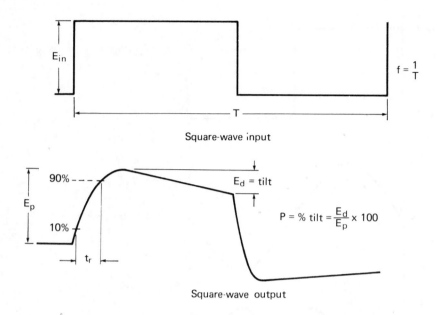

Square-wave input

Square-wave output

Figure 5.30 Typical square-wave output from an RC amplifier.

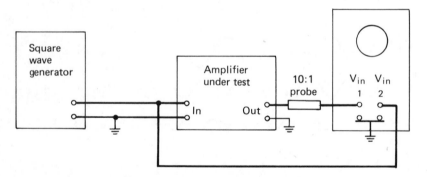

Figure 5.31 Test setup for square-wave testing of an amplifier.

output square waves. Since an amplifier saturating may produce a square wave which looks ideal, the square-wave amplitude is to be doubled to verify that the output response doubles in amplitude without a change of shape. The PRF is to be adjusted from a minimum equal to the f_{low} of the amplifier to at least one ninth of its high-frequency response. The response of the amplifier is determined by comparing the output waveform to the typical patterns shown in Fig. 5.32.

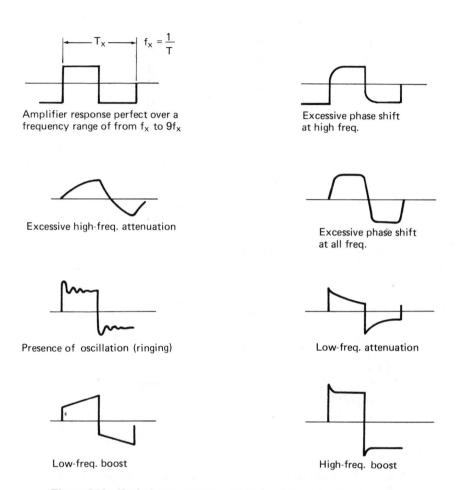

Figure 5.32 Typical square-wave patterns for determining the characteristics of an amplifier.

5.19   PULSE MEASUREMENTS

With a square wave having a majority of its energy at the low end of the spectrum, high-frequency amplifiers and digital switching circuits are usually pulse (impulse) tested instead. The duration of the pulse should be short compared with its period, with the narrower the pulse, the higher its frequency spectrum. Refer to Figs. 5.33(a) and (b) for a typical pulse waveform.

The important parameters of the output pulse to be measured are (1)

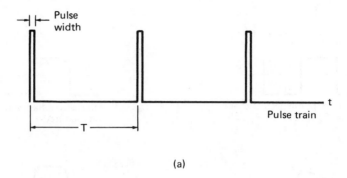

(a)

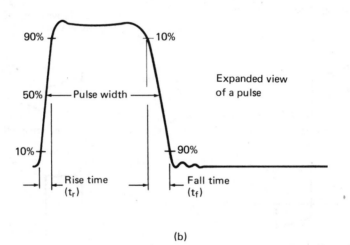

(b)

Figure 5.33   Waveform of a typical pulse train, with an expansion of the pulse.

PRF, (2) duty cycle, (3) pulse width, (4) rise time, and (5) fall time. For a more detailed discussion of pulse parameters, refer to Chapter 12.

The test setup for pulse testing a circuit is shown in Fig. 5.34. With the scope set to ext. trigger, the vertical sens. and sweep speed controls are adjusted for a stationary display of the pulse output. Refer to Figs. 5.33(a) and (b). Measure the output pulse amplitude, pulse width, and pulse period T. The pulse repetition frequency and duty cycle are

$$\text{PRF} = \frac{1}{T}$$

$$\text{duty cycle} = \frac{\text{pulse width}}{T} = (\text{pulse width}) \times \text{PRF}$$

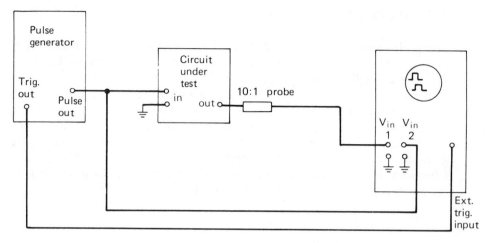

Figure 5.34   Test setup for performing pulse measurements.

The CRO sweep speed and sweep magnifier controls (if required) are now adjusted such that the pulse occupies [Fig. 5.33(b)] a majority of the screen. Measure the rise time t_r (10% to 90% Pt.) of the output pulse. With many circuits having a rise time on the order of the scope rise-time capabilities, the error caused by the oscilloscope and pulse generator may be determined by

$$t_{r(\text{dev})} = \sqrt{t_{r(\text{meas})}^2 - t_{r(\text{scope})}^2 - t_{r(\text{PG})}^2}$$

where $t_{r(\text{dev})}$ = actual rise time of device under test,

  $t_{r(\text{meas})}$ = overall displayed rise time (scope and signal),

  $t_{r(\text{scope})}$ = scope rise time,

  $t_{r(\text{PG})}$ = pulse generator rise time.

An illustrative example will explain this measurement correction.

Example

A pulse generator with a t_r of 2 ns is utilized in determining the rise time of an amplifier. If a 100-MHz scope measured a t_r of 10 ns, what is the actual rise time of the amplifier?

$$t_{r(\text{scope})} = \frac{0.35}{f_h} = \frac{0.35}{\text{BW}} = \frac{0.35}{100 \times 10^6} = 0.0035 \ \mu s = 3.5 \ \text{ns}$$

$$t_{r(\text{dev})} = \sqrt{t_{r(\text{meas})}^2 - t_{r(\text{scope})}^2 - t_{r(\text{PG})}^2} = \sqrt{10^2 - 3.5^2 - 2^2}$$

$$= \sqrt{100 - 12.25 - 4}$$

$$= 9.15 \ \text{ns}$$

To measure the fall time accurately, a scope with a delay feature may be necessary. Refer to Section 5.9.

5.20 RECEIVER ALIGNMENT

Sweep alignment of an IF amplifier, an FM detector, or associated receiver circuits may be accomplished with a sweep generator and an oscilloscope. The sweep oscillator IF output is loosely coupled into circuit under test, with a scope and a demodulator probe monitoring the output. Refer to Fig. 5.35. With the scope operated as a X-Y recorder (see Section 5.6), the exact IF response of the amplifier will be traced on the face of the oscilloscope. If the sweep oscillator hasn't a marker generator output, an external signal generator maybe utilized for calibrating the exact frequency for proper alignment. Refer to Chapters 12 and 13 for a more detailed discussion of receiver alignment utilizing the scope and other measuring instruments.

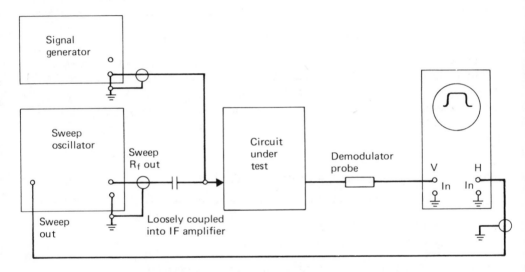

Figure 5.35   Test setup for alignment of an IF amplifier.

5.21 SUMMARY

The most versatile and most frequently used instrument for performing electronic measurements is the oscilloscope. Its basic operation has been presented with stress placed on such features as dual trace, sweep synchronization, magnification, and delay.

The various types of probes were discussed with a procedure for calibrating and compensating the attenuator probe.

With an understanding of scope operation, basic measurements of voltage, frequency, and phase were presented, proceeding to square waves and pulse testing and concluding with the alignment of a receiver IF circuit.

REVIEW QUESTIONS

1. Explain what the three positions (internal, external, line) of the trigger source control accomplish.

2. The sweep speed of a scope is set to 1 μs/div., with the sweep magnification set to ×5. A complete cycle of a sinusoidal voltage occupied 8.5 divisions. Determine the frequency.

3. Four complete periods of a 1,000-Hz sinusoidal signal occupied 6 div. Determine the sweep speed (seconds per division) of the oscilloscope.

4. Explain the function of the trigger slope and trigger level controls.

5. What are the differences between the chop and alternate modes of a dual trace scope?

6. What is an active probe, and what are its advantages over a compensated attenuator?

7. The sweep speed is set to 1 ms/div., with the vertical sens. set to 10 mV/div. Sketch the waveform patterns of the following signals:
   a. Square wave whose frequency is 500 Hz, 40 mV peak to peak.
   b. Sinusoidal voltage of 50 mV rms with a frequency of 2,000 Hz.

8. Discuss two methods of measuring phase utilizing an oscilloscope.

9. An RC amplifier undergoing a square-wave test had a tilt of 15%. If the PRF of the square wave is 100 Hz, determine the approximate low-frequency response of the amplifier.

10. The output pulse rise time of a video amplifier was measured as 0.5 μs. What is the approximate bandwidth of the amplifier?

11. With the scope set to alternate, with a sweep speed of 50 ms/cm, sinusoidal waveform a lags sinusoid b by 2 cm. If a complete cycle of a sinusoid is 8 cm, determine the phase relationship between a and b in degrees.

12. The Lissajous pattern of two sinusoids are shown in Figs. 5.36(a) and (b). Determine the phase relationship between the two waveforms.

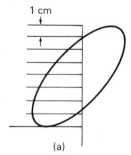

1 cm

(a)

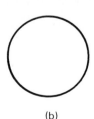

(b)

Figure 5.36

13. The scope is connected as an X-Y recorder to determine the frequency
     by the Lissajous pattern method. If the sinusoid into the vertical
     amplifier is 4,000 Hz, determine the frequency of sinusoidal input at
     the horizontal input. Refer to Fig. 5.37 for the Lissajous pattern.

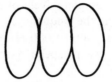

Figure 5.37

14. A sense resistor of 10 Ω is utilized to measure the current flowing in a
     pulse circuit. The vertical attenuator is set to 5 V/cm, with the resultant
     waveform shown in Fig. 5.38. Determine the peak pulse current.

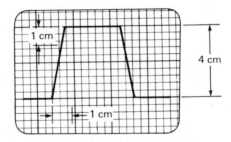

Figure 5.38

15. A scope sweep speed is set to 0.5 μs/cm, with the resultant waveform
     shown in Fig. 5.38. Find
     a.   The rise time, t_r.
     b.   The pulse width.
     c.   The fall time.

16. The pulse waveform shown in Fig. 5.38 is measured with a scope with
     a vertical amplifier bandwidth (BW) of 10 MHz. Determine the true
     rise time of the waveform. *Hint*: $t_r = 0.35/\text{BW}$.

6
DISCRETE COMPONENT MEASUREMENTS

6.1 INTRODUCTION

There are three basic components utilized in electronic equipment: resistors, capacitors, and inductors.

In actuality, the three components are complex, having all three of the characteristics of resistance, capacitance, and inductance. How a particular component behaves in a circuit depends on its biasing voltage and frequency of operation. As an example, a simple inductor operating at a frequency greater than its self-resonance frequency will act as a capacitor. A wire-wound resistor has an intrinsic inductance which at high frequencies will cause the component to act as an inductor.

An electronic measurement which will give us an indication of these complex characteristics is determining the quality factor (Q) or dissipation factor (D) in addition to the component's basic resistance, capacitance, or inductance.

In this chapter we shall formulate several different measurement techniques utilizing basic laboratory equipment—sophisticated bridges, impedance meters, and Q meters—for determining a component's complex characteristics.

6.2 IMPEDANCE CONCEPTS

With a voltage (ac or dc) impressed across a two-terminal passive device, a current will flow, with the ratio of voltage to current being defined as impedance. When the impressed voltage is dc the resultant ratio of voltage to current (E/I) is defined as the dc resistance, R_{dc}:

$$R_{dc} = \frac{E}{I}$$

For a sinusoidal (ac) input voltage, the ratio E/I is generally a complex term referred to as impedance (Z). This impedance in Cartesian form is given by

$$Z = \left(\frac{E}{I}\right)_{ac} = R_{ac} + jX$$

The real part of the impedance (R_{ac}) is referred to as the *effective* or *ac resistance*. This resistance (R_{ac}) is the element in an ac curcuit which is dissipative, with the spent power equal to $I^2 R_{ac}$. The imaginary part (X) is called reactance and represents the energy storage of the impedance. The quantities R_{ac} and X both vary with frequency. At dc (zero frequecy) the reactance X is either zero or infinite in value, depending on the component.

Many times the reciprocal of impedance is required. The reciprocal of a dc resistance is called conductance, G:

$$G_{dc} = \frac{1}{R_{dc}}$$

The reciprocal of the ac impedance Z is called admittance, Y, where

$$Y = \frac{1}{Z}$$

Since the ac impedance (Z) is a complex value consisting of a real and imaginary part, the reciprocal of Z will also be complex, with the real part called the conductance, G, and the imaginary part called the susceptance, B.
With $Z = R_{ac} + jX$,

$$\frac{1}{Z} = \frac{1}{R + jX} = Y = G_{ac} + jB$$

$$Y = \frac{1}{Z} = G_{ac} + jB$$

Multiplying and dividing Y by its complex conjugate, $Y^* = 1/(R - jX)$, we obtain

$$\frac{Y}{1} \times \frac{Y^*}{Y^*} = \frac{1}{R + jx} x \frac{R - jX}{R - jX} = \frac{R - jX}{R^2 + X^2} = G_{ac} + jB$$

Therefore,

$$G_{ac} = \frac{R}{R^2 + X^2} \qquad |B| = \frac{X}{R^2 + X^2}$$

Note that G and B both vary with frequency.

The three basic elements—resistors, inductors, and capacitors—depicted schematically in Fig. 6.1 are ideal components. The actual physical components are highly complex, with each exhibiting all three characteristics in varying degrees.

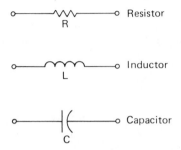

Figure 6.1   Ideal resistor, inductor, and capacitor.

Refer to Figs. 6.2(a), (b), and (c) for the equivalent circuits of an actual physical resistor, inductor, and capacitor.

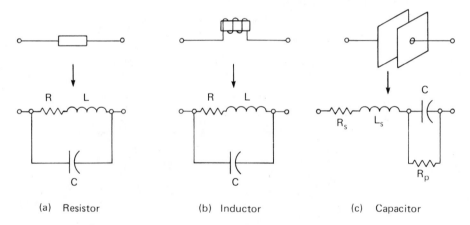

Figure 6.2   Equivalent circuits of physical resistor, inductor, and capacitor.

To determine how a physical component will act in a circuit, it is important to determine the ratio of its resistance to reactance or vice versa. A component operated in a particular frequency range will have a complex impedance consisting of a real and imaginary part, which may be inductive or capacitive.

For inductive reactive components an important characteristic to be measured is the ratio of its reactance to its resistance, called the storage factor Q.

It was previously stated that a component's reactance is an indication of its energy storage, while its real component is a measure of its dissipation of power loss. The storage factor is therefore a measure of a component's stored to lost energy.

The reciprocal of Q, called the dissipation factor, D, is another important characteristic and is usually determined for components which are capacitive in nature:

$$D = \frac{1}{Q}$$

At any one frequency a complex impedance may be simulated by two ideal circuit elements, an equivalent resistance and an equivalent capacitance or inductance. The equivalent impedance can be assumed to consist of elements connected in series, as depicted in Fig. 6.3(a) For the inductive circuit of Fig. 6.3(a), Q and D may be calculated by

$$Q = \frac{X}{R} = \frac{\omega L_s}{R_s}$$

where ω = radian frequency, with

$$D = \frac{1}{Q}$$

$$= \frac{R_s}{\omega L_s}$$

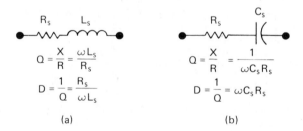

(a)   (b)

Figure 6.3   (a) Resistor in series with an inductor. (b) Resistor in series with a capacitor.

For a capacitive series circuit as shown in Fig. 6.3(b),

$$Q = \frac{X}{R} = \frac{1}{\omega C_s R_s}$$

$$D = \frac{1}{Q} = \omega C_s R_s$$

The complex impedance may also be simulated by the parallel combination of two ideal elements. For an inductive circuit, as shown in Fig. 6.4(a),

$$Q = \frac{R_p}{\omega L_p}$$

$$D = \frac{1}{Q} = \frac{\omega L_p}{R_p}$$

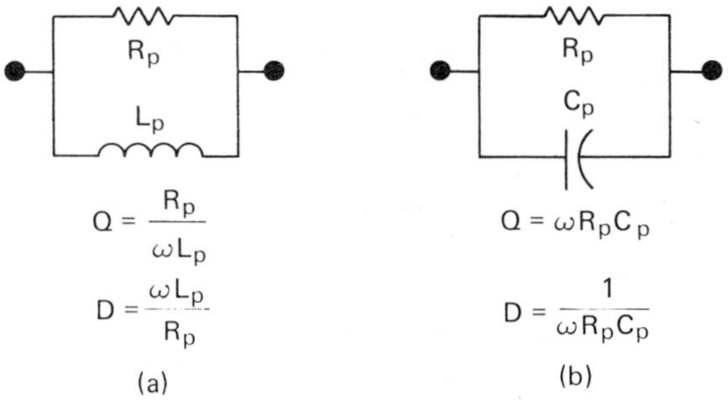

$$Q = \frac{R_p}{\omega L_p}$$

$$D = \frac{\omega L_p}{R_p}$$

$$Q = \omega R_p C_p$$

$$D = \frac{1}{\omega R_p C_p}$$

(a)   (b)

Figure 6.4 (a) Resistor in parallel with an inductor. (b) Resistor in parallel with a capacitor.

For a capacitive parallel circuit, as shown in Fig. 6.4(b),

$$Q = \omega R_p C_p$$

$$D = \frac{1}{Q} = \frac{1}{\omega R_p C_p}$$

For large-value capacitors (electrolytic capacitors) a term called the power factor (F_p) is usually measured instead of the dissipation factor D. The term power factor is frequently used interchangeably with the term dissipation factor. For a storage factor (Q) greater than or equal to 10, the power factor is approximately equal to the dissipation factor:

$$F_p = \cos \theta = D \qquad \text{for } Q \gtrless 10$$

The proof follows: Referring to circuit of Fig. 6.5,

$$I = \frac{V}{Z}; \qquad P = IV \cos \theta = I^2 R$$

$$P_{\text{lost}} = IV \cos \theta = I^2 R$$

$$F_p = \cos \theta = \frac{P_{\text{lost}}}{IV}$$

$$\cos \theta = \frac{P_{\text{lost}}}{IV} = \frac{I^2 R}{IV} = \frac{I^2 R}{I^2 Z} = \frac{I^2 R}{I^2 Z} = \frac{R}{Z}$$

$$\cos \theta = \frac{R}{Z} = \frac{R}{(R^2 + X^2)^{1/2}}$$

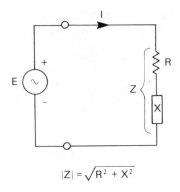

$$|Z| = \sqrt{R^2 + X^2}$$

Figure 6.5

for $Q \gtrless 10$, $X \gtrless 10R$;

$$\therefore \quad R^2 + X^2 \approx X^2$$

$$F_p = \frac{R}{\sqrt{X^2}} = \frac{R}{X} = D$$

6.3 RESISTANCE MEASUREMENT (LOW FREQUENCY)

6.3.1 Volt-Ammeter Method

A simple method of measuring the dc resistance of any component is simply to connect the unknown component in a dc circuit and measure the current and voltage. See Fig. 6.6.

The ratio V/I is the dc resistance of the device. The accuracy of this measurement depends on the calibration, the stability of the two meters, and the loading effect of the voltmeter.

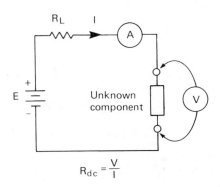

$$R_{dc} = \frac{V}{I}$$

Figure 6.6 Simple method of measuring the *dc* resistance of any component.

6.3.2   Wheatstone Bridge Method

A more accurate method is to connect the unknown component in a bridge circuit; when balanced, the unknown value is determined by comparing it against a standard known resistor.

The Wheatstone bridge is such a circuit, depicted in Fig. 6.7 and discussed in Chapter 2. The detector may consist of a galvanometer for dc measure-

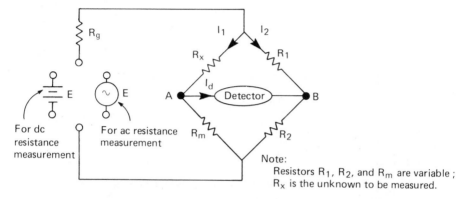

Figure 6.7   Wheatstone bridge. *Note:* resistors R_1, R_2, and R_m are variable; R_x is the unknown to be measured.

ments or any ultrasensitive voltage measuring device, such as a dc coupled oscilloscope which can measure voltages in the millivolt range. The resistors R_1, R_2, and R_m are adjusted until the detector reads zero, which indicates a balanced (or nulled) condition. With the bridge balanced, zero current flows through the detector, and zero voltage is across it. Refer to Chapter 2 for a detailed analysis of a Wheatstone bridge.

As derived in Chapter 2, for a balanced Wheatstone bridge, the unknown resistor R_x may be determined by

$$R_x = \frac{R_m R_1}{R_2}$$

With R_m, R_1, and R_2 being standard known resistors, the unknown resistance R_x may be determined. In a commercial Wheatstone bridge R_m is the multiplier resistor and R_1 and R_2 are calibrated potentiometers. When the bridge is balanced, R_x is determined by reading the dial position corresponding to the values of R_m, R_1, and R_2. Refer to Fig. 6.8 for a simplified commercial Wheatstone bridge. Note that the unknown resistance R_x measurement is independent of the voltage source or detector impedance when the bridge is balanced. The Wheatstone bridge is capable of dc or ac resistance measurements by the utilization of a corresponding dc or ac voltage source.

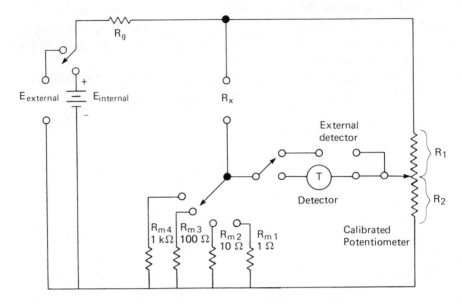

Figure 6.8   Simplified commercial Wheatstone bridge.

6.4   AC IMPEDANCE MEASUREMENTS (LINE FREQUENCY POWER) VIA AMMETER-VOLTMETER-WATTMETER METHOD

The measurements which follow are for components which operate in ac line power circuits such as balasts, motor starting capacitors, or equivalent circuits of a power transformer. When performing ac impedance measurements, one must determine the ac resistance (R_{ac}) plus the reactance (X). A circuit which can be utilized in performing these measurements is shown in Fig. 6.9. The wattmeter measures the power lost in the component, which is equal to $I^2 R_{ac}$. The ammeter and voltmeter determine the current (I) and the voltage (V) across the unknown component. The component impedance is found by

$$Z = \frac{V}{I} = R_{ac} + jX$$

The component's resistance, R_{ac}, is the wattmeter reading divided by its current squared:

$$R_{ac} = \frac{P}{I^2}$$

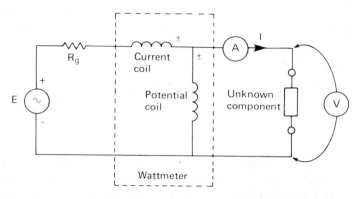

Figure 6.9 Meter setup to measure the impedance, resistance, and reactance of an unknown component. *Note:* all meters are *ac*, with the internal wattmeter dissipation assumed to be negligible or subtracted from the wattmeter reading for better accuracy.

With Z and R_{ac} known, the reactance is determined by

$$Z = \sqrt{R_{ac}^2 + X^2}$$

$$Z^2 = R_{ac}^2 + X^2; \qquad X^2 = Z^2 - R_{ac}^2$$

$$X = \sqrt{Z^2 - R_{ac}^2}$$

Since the frequency (f) of the generator is known, the particular inductance or capacitance may be determined by

$$L = \frac{X_L}{2\pi f}$$

$$C = \frac{1}{X_C 2\pi f}$$

6.5 LOW-FREQUENCY AC CAPACITOR MEASUREMENTS

6.5.1 Capacitor Leakage Measurements

An ideal capacitor, fully charged, with rated operating dc voltage across it will theoretically have zero dc current flowing through it. In actuality there is a minute leakage current through the capacitor. The ratio of operating dc voltage to leakage current is called capacitor leakage resistance.

A capacitor with an abnormally low leakage resistance when compared to other capacitors of a similar type is to be considered defective. See Figs. 6.10(a), (b), and (c). Note that, depending on the type of capacitor, the leakage resistance can vary greatly. A capacitor made of mica, ceramic, paper,

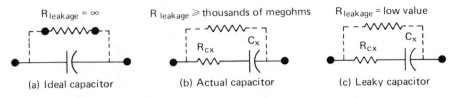

$R_{leakage} = \infty$	$R_{leakage} \geqslant$ thousands of megohms	$R_{leakage} =$ low value
(a) Ideal capacitor	(b) Actual capacitor	(c) Leaky capacitor

Figure 6.10 Whether or not a capacitor is called leaky and is rejected depends on its type. A glass capacitor may have a typical $R_{leakage}$ of 10^9 Ω, while typical electrolytic capacitors have an $R_{leakage}$ of 500 kΩ.

or glass may have a leakage resistance on the order of thousands of megohms, while the normal leakage resistance of a good electrolytic capacitor may be less than 1 MΩ. It is therefore good measurement procedure to test a capacitor for leakage prior to actually measuring its capacitance value, since excessive leakage will cause an error in capacitance measurements.

The leakage resistance of a capacitor is measured with rated (dc operating) voltage across it (which may be as high as 500 V). A simple circuit which can measure the leakage resistance of a capacitor is shown in Fig. 6.11.

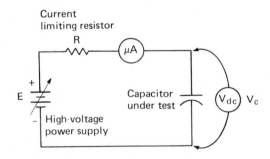

Figure 6.11 Simple circuit to measure leakage resistance of a capacitor.

The leakage resistance is the ratio of V_c to I. A commercial Wheatstone bridge may also be utilized to measure leakage resistance if the internal low-voltage dc source is replaced with a high-voltage source (with the capacitor to be tested connected across the R_x terminals). See Fig. 6.8.

6.5.2   ac Bridge Capacitor Measurements

The capacitance, dissipation factor (D), and power factor (F_p) of a capacitor may be measured with a modified Wheatstone bridge circuit as shown in Fig. 6.12. The exciting voltage can be 60 hz ac voltage or a 1 khz audio signal. Detection can be accomplished with a high-gain oscilloscope, a sensitive ac voltmeter, or a pair of earphones.

At low frequencies the complex circuit of a capacitor is simplified to a pure capacitor in series with a small resistor. See Fig. 6.13. To balance the

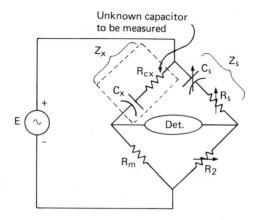

Figure 6.12 Capacitor bridge circuit for measuring capacitors with a series resistance.

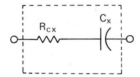

Figure 6.13 Low-frequency equivalent circuit of a capacitor.

bridge with a complex component, the bridge must also be complex in nature.

For the bridge to be balanced, two balance conditions must be satisfied, one for the resistive (real) and one for the reactive (imaginary) component. With reference to Fig. 6.12 and with the circuit in balance, $Z_x R_z = Z_s R_m$ and

$$Z_x = \frac{Z_s R_m}{R_z}$$

With $Z_x = [R_{cx} + (1/j\omega c_x)]$ and $Z_s = [R_s + (1/j\omega c_s)]$,

$$R_{cx} + \frac{1}{j\omega c_x} = \left(R_s + \frac{1}{j\omega c_s}\right)\left(\frac{R_m}{R_z}\right)$$

$$R_{cx} + \frac{1}{j\omega c_x} = \frac{R_s R_m}{R_z} + \frac{R_m}{R_z j\omega c_s}$$

For the above equation to be satisfied (real parts and imaginary parts equal to each other),

$$R_{cx} = \frac{R_s R_m}{R_z} \qquad \frac{1}{j\omega c_x} = \frac{R_m}{R_z j\omega c_s}$$

$$R_{cx} = \frac{R_s R_m}{R_z} \qquad C_x = \frac{C_s R_z}{R_m}$$

In commercial capacitor bridges, R_s is calibrated in units of dissipation factor (D) or in units of power factor (F_p), which technicians are more concerned with.

With a knowledge of the frequency and capacitance value R_{cx} is detemined by

$$D = \frac{1}{Q} = \omega C_x R_{cx}$$

$$R_{cx} = \frac{D}{\omega C_x} = \frac{D}{2\pi f C_x}$$

For small mica, ceramic, paper, glass, etc., capacitors which have a very small dissipation factor ($R_{cx} \approx 0$), it follows that the storage factor $Q = 1/D > 10$. Therefore,

$$F_p = \cos \theta = D = \omega C_x R_{cx}$$

The measurement of an electrolytic capacitor which has a normally large series resistance (large D) and whose capacitance is a function of a polarizing (dc) voltage is accomplished with the modified bridge circuit shown in Fig. 6.14.

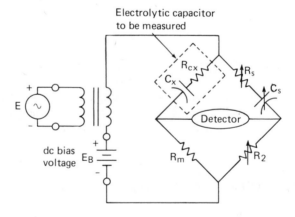

Figure 6.14 Capacitance measuring bridge with external *dc* bias for electrolytic capacitor.

In electrolytic capacitor bridges the variable resistor R_s is usually calibrated in units of power factor (F_p).

The capacitance bridge circuit of Fig. 6.12 is ineffective in performing accurate measurements on capacitors with extremely low dissipation factors. This is due to the practical limitations of fabricating a standard capacitor (C_s) which has a dissipation factor (D) considerably lower than the capacitor to be tested. To overcome this problem the *Schering bridge*, depicted in Fig.

6.15, may be utilized. With the bridge balanced,

$$Z_x Z_2 = R_1 Z_s$$

$$Z_x = \frac{R_1 Z_s}{Z_2}$$

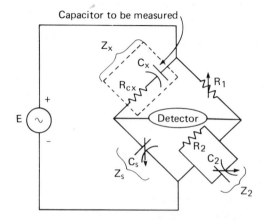

Figure 6.15   Schering bridge utilized for capacitors with a very low dissipation factor D.

With $Z_x = [R_{cx} + (1/j\omega c_x)]$, $Z_2 = R_2/(1 + j\omega R_2 C_2)$, and $Z_s = 1/j\omega c_s$,

$$R_{cx} + \frac{1}{j\omega c_x} = \frac{(R_1)(1 + j\omega R_2 C_2)}{(j\omega c_s)R_2}$$

$$= \frac{R_1}{R_2 j\omega c_s} + \frac{R_1 j\omega R_2 C_2}{j\omega c_s R_2}$$

or

$$R_{cx} = \frac{R_1 C_2}{C_s} \qquad C_x = \frac{R_2 C_s}{R_1}$$

The commercial Schering bridge will have its dials calibrated in terms of C_s and D instead of R_{cx}. With a knowledge of C_s and D, R_{cx} may be calculated as

$$R_{cx} = \frac{D}{\omega c_s} = \frac{D}{2\pi f c_s}$$

where f = frequency of the bridge signal.

6.6   INDUCTANCE BRIDGE MEASUREMENTS

The equivalent circuit of an actual inductor may be inductive or capacitive depending on its self-resonant frequency. See Figs. 6.16(a), (b), and (c). A coil tested with a frequency equal to its resonant frequency will have an impedance much greater than its true inductive reactance, resulting in a

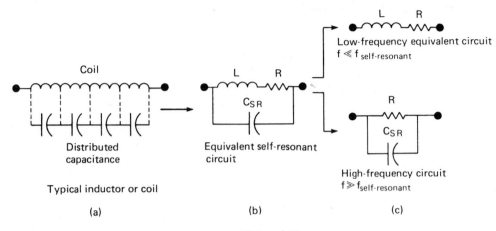

Figure 6.16

large measurement error. To overcome this problem, coils should be tested
at a frequency much lower than the resonant frequency. The bridge circuit of
Fig. 6.12 may be modified to measure inductances with the standard R_s and
C_s replaced with an L_s and R_s. Refer to the bridge circuit of Fig. 6.17. At
balance, the same impedance calculations as for the capacitor apply; therefore
it can be shown that

$$R_x = \frac{R_s R_m}{R_2} \qquad L_x = \frac{L_s R_m}{R_2}$$

The bridge circuit of Fig. 6.17 is usually not used due to the requirements
of an expensive stable and accurate standard inductor L_s. The circuit may be
modified using a standard capacitor. This circuit configuration is called a
Maxwell bridge, shown in Fig. 6.18.

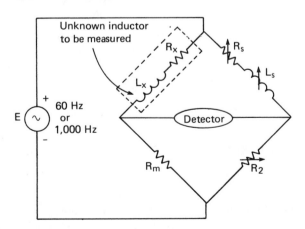

Figure 6.17   Basic bridge for inductance measurements.

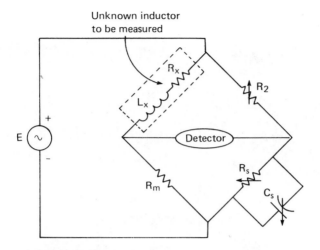

Figure 6.18   Maxwell bridge for inductance measurements.

Since the capacitance draws a leading current while the inductive current lags, the reactive components are located opposite each other to obtain phase cancellation. It can be shown that the balance equations are

$$R_x = \frac{R_m R_2}{R_s} \qquad L_x = R_2 R_m C_s$$

As with the capacitor bridges, the R_x of the inductor is disregarded in favor of the storage factor Q. The resistor R_x of commercial bridges is calibrated in Q, where

$$Q = \frac{\omega L_x}{R_x} = \omega C_s R_x$$

Since the Q measurement is a function of ω, the oscillator frequency should be stable with a good sinusoidal waveform.

6.7 IMPEDANCE BRIDGES

An impedance bridge is a single instrument capable of measuring resistance, inductance, and capacitance with their associated dissipation and quality factors. See Fig. 6.19 for an automatic digital *LRC* meter. By adjusting various controls and switches, the impedance bridge can become a Wheatstone bridge for measuring (dc or ac) resistance or a capacitor, inductor bridge. The impedance bridge doesn't actually measure a component's impedance (Z) but rather its capacitance and dissipation factor or its inductance and quality factor. Most impedance bridges operate with an excitation voltage of dc or 1,000 Hz ac. To determine accurately a component's characteristic at a frequency other than 1 kHz may be accomplished by operating the bridge with an external generator. For example, an inductive component's charac-

Figure 6.19 Automatic digital *LRC* meter courtesy of Hewlett-Packard.

teristic operating at 100 kHz is required. The impedance bridge is externally excited with a 100-kHz generator, and at bridge balance L_x and Q are read off the dials. The ac resistance of the inductor at 100 kHz is determined by

$$Q = \frac{\omega L_x}{R_x}$$

or

$$R_x = \frac{\omega L_x}{Q} = \frac{2\pi f L_x}{Q}$$

where $f = 100$ kHz

6.8   HIGH-FREQUENCY COMPONENT MEASUREMENTS UTILIZING THE Q METER

An important application of the inductor is in a resonant circuit acting as a band-pass filter. See Fig. 6.20a. The bandwidth (at its 3-dB points) is determined by

$$\text{BW} = \frac{f_0}{Q}$$

where f_0 is the resonant frequency and Q the quality factor. See Fig. 6.20(b) and (c) for the bandwidth of a high- and low-Q resonant circuit.

As previously mentioned, the Q of a coil varies inversely with its ac resis-

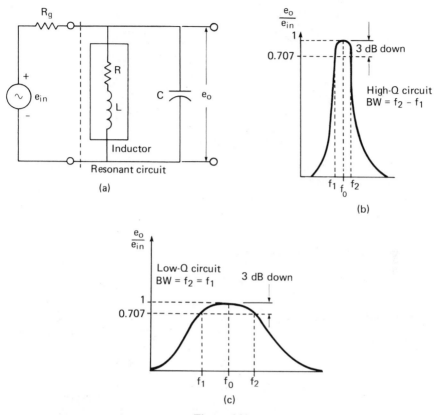

Figure 6.20

tance (R_{ac}), which in turn varies with frequency. This resistance variation with frequency is due to skin effect. At high frequencies, the current in a conductor is redistributed, with the least concentration at the center and with maximum concentration at the conductor. See Fig. 6.21. Since the resistance increases as the conducting area decreases, a high frequency will cause an increase in component resistance.

An instrument widely used in performing accurate Q measurements is the Q meter; its basic circuit is shown in Fig. 6.22(a). This instrument is superior to the impedance bridge in that the Q measurement is made at the operating frequency of the coil. Typical Q meters operate over a frequency range of from 1 kHz to 300 MHz. See Fig. 6.22(b) for a commercial Q meter manufactured by Hewlett-Packard.

The circuit operates as follows: When connected to the Q meter the coil to be measured forms a simple series circuit. The capacitor (C) is varied until V_c is a maximum, which occurs when the circuit is at resonance. At resonance $V_c = Qe_{in}$, and with e_{in} being a fixed known value, the amplitude of V_c can be directly calibrated to read Q. It is left as an exercise for the student to

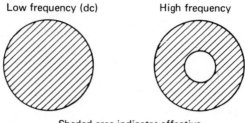

Low frequency (dc)   High frequency

Shaded area indicates effective
current conductive area

Figure 6.21 Effective cross-sectional area of a conductor for current flow showing skin effect.

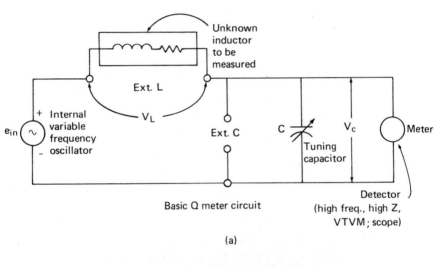

(a)

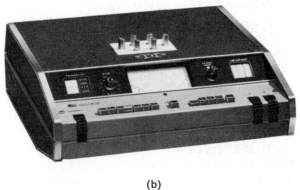

(b)

Figure 6.22 (b) courtesy of Hewlett-Packard.

prove that for a series resonant circuit $V_c = V_L = Qe_{in}$. After determining the capacitor reading, we can calculate the inductance L_x by

$$\omega_0 L = \frac{1}{\omega_0 c} \qquad \text{(system at resonance)}$$

$$L = \frac{1}{\omega_0^2 C} = \frac{1}{(2\pi f_0)^2 C}$$

f_0 is the frequency of the Q meter internal oscillator.

The inductor series resistance R is determined by

$$Q = \frac{2\pi f_0 L}{R} = \frac{1}{2\pi f_0 RC}$$

$$R = \frac{1}{2\pi f_0 CQ}$$

where C is the Q meter tuning capacitor value.

6.8.1   Measurement of an Inductor's Self-Resonant Capacitance

Associated with every inductor is a self-resonant capacitance, as shown in Figs. 6.16(a) and (b). The Q meter can determine this value by the following simple method. The inductor is connected to the Q meter with the (Q meter) tuning capacitor set to its maximum capacitance, C_{max}. The meter oscillator frequency (f_r) is then adjusted to obtain resonance. The oscillator frequency is set to $2f_r$, and the tuning capacitor is adjusted to a new value C_2 until resonance is restored. The self-resonant capacitance of the inductor is then given by

$$C_{SR} = \frac{C_{max} - 4C_2}{3}$$

6.8.2   High-Frequency Capacitor Measurements

The physical capacitor operating at a high frequency may be represented by the equivalent circuit shown in Fig. 6.23(a). To assess the capacitor in a particular circuit at a particular frequency a knowledge of its effective capacitance (C_x) is more useful in a majority of cases. See Fig. 6.23(b). The effective capacitance (C_x) is made up of the capacitor and leads inductance (L_w).

The Q meter can be utilized to measure the effective capacitance (C_x) and the shunt resistance (R_{xp}) by the following procedure. With reference to Fig. 6.24, a suitable self-resonating inductor (supplied with the Q meter) for the proper frequency range is connected across the "Ext. L" connections. The Q meter tuning capacitor is set close to its maximum with the setting (C_1) noted. The frequency (f_1) is then adjusted for resonance, noting its value and the Q of coil Q_1. The capacitance to be measured is then connected across the "Ext. C" connections, and the tuning capacitor is readjusted for resonance. The new tuning capacitor value (C_2) and the Q of the new circuit (Q_2) are

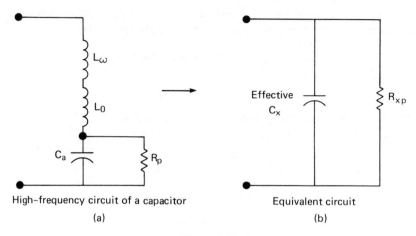

High-frequency circuit of a capacitor
(a)

Equivalent circuit
(b)

Figure 6.23

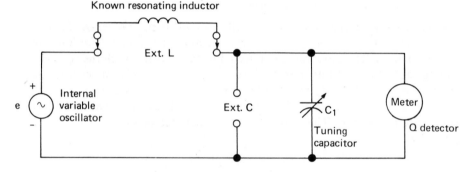

Figure 6.24 Q meter circuit for measuring effective capacitance.

noted. The following terms may be determined:

1. Unknown effective capacitor value C_x:
$$C_x = C_1 - C_2$$

2. Dissipation factor D:
$$D = \frac{Q_1 - Q_2}{Q_1 Q_2} \frac{C_1}{C_1 - C_2}$$

3. Shunt resistance of capacitor R_{xp}
$$R_{xp} = \frac{Q_1 Q_2}{Q_1 - Q_2} \frac{1}{2\pi f_1 C_1}$$

6.8.3   High-Frequency Resistance Measurements

Similar to the capacitor, the resistor operating at a high frequency is a complex circuit. See Fig. 6.25(a). The complex circuit may be represented by the equivalent circuit shown in Fig. 6.25(b). The equivalent circuit parameters

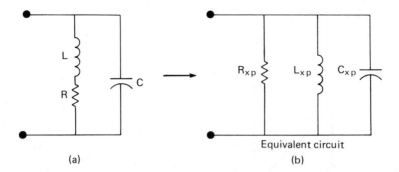

(a)   (b)

Figure 6.25   High-frequency circuit of a resistor with its equivalent circuit.

may be measured with the Q meter as follows: Depending on the frequency
range in which the resistor parameters are to be determined, a self-resonating
inductor is connected across the "Ext. L" terminals. See Fig. 6.24. With the
Q meter frequency (f_0) properly set, the tuning capacitor (C_1) is adjusted
for resonance, with its value (C_1), f_0, and the Q of the circuit (Q_1) noted.
The resistor to be measured is connected across the "Ext. C" terminals, and
the tuning capacitor is readjusted for resonance. The new tuning capacitor
setting (C_2) and the new Q of circuit (Q_2) are noted.

The equivalent resistor parameters are determined by the following
equations:

$$R_{xp} = \frac{Q_1 Q_2}{Q_1 - Q_2} \frac{1}{2\pi f_0 C_1}$$

$$C_{xp} = C_1 - C_2$$

$$L_{xp} = \frac{1}{(2\pi f_0)^2 C_{xp}}$$

6.9   SUMMARY

In this chapter we have covered basic component measurements determining
the following: resistance, impedance, reactance, quality factor, dissipation
factor, and power factor.

Low-frequency measurements were performed utilizing basic meters and
bridges such as the Wheatstone, capacitor, Schering, Maxwell, and impedance bridges. High-frequency component measurements were then discussed
utilizing the Q meter.

For realistic component measurements, the circuit or bridge used to
perform the measurement should have a frequency capability simulating the
actual frequency the component is to be subjected to.

Figure 6.26 shows the formulas used in bridge measurements, summarized
for reference.

Wien bridge

$$\frac{C_x}{C_s} = \frac{R_b}{R_a} - \frac{R_s}{R_z}$$

$$C_s C_x = 1/\omega^2 R_s R_z$$

For measurement of frequency, or in a frequency-selective application, if

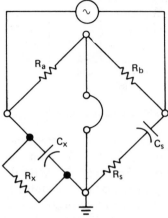

we make $C_x = C_s$, $R_x = R_s$, and $R_b = 2R_a$, then

$$f = \frac{1}{2\pi C_s R_s}$$

(a)

Schering bridge

$$C_x = C_s R_b/R_a$$

$$1/Q_x = \omega C_x R_x = \omega C_b R_b$$

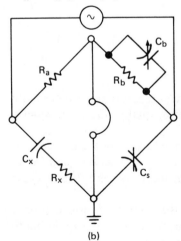

(b)

Measurement with capacitor is series with unknown

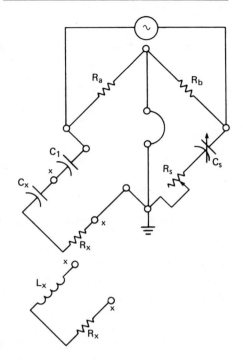

When $C''_s > C'_s$,

$$L_x = \frac{1}{\omega^2} \frac{R_a}{R_b C'_s} \left(1 - \frac{C'_s}{C''_s} \right)$$

Initial balance (unknown terminals x – x short circuited):

C'_s and R'_s

Final balance (x – x un-shorted):

C''_s and R''_s

Then the series resistance is

$$R_x = (R''_s - R'_s) R_a/R_b$$

$$C_x = \frac{R_b C'_s C''_s}{R_a (C'_s - C''_s)}$$

$$= \frac{R_b}{R_a} C'_s \left(\frac{C'_s}{C'_s - C''_s} \right) - 1$$

(c)

Figure 6.26 Bridge measurement formulas.

Measurement of direct capacitance

Connection of N to N' places C_{np} across phones, and C_{np} across R_b which requires only a small readjustment of R_s.

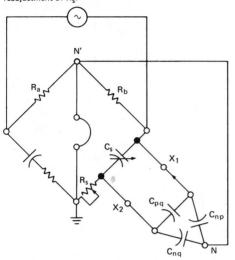

Initial balance' Lead from P disconnected from X_1 but lying as close to connected position as practical.

Final balance: Lead connected to X_1.

By the substitution method above,
$C_{pq} = C'_s - C''_s$

(d)

Hay bridge

For measurement of large inductance.

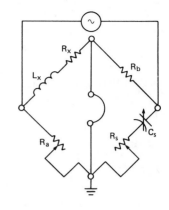

$$L_x = \frac{R_a R_b C_s}{1 + \omega^2 C_s^2 R_s^2}$$

$$Q_x = \frac{\omega L_x}{R_x} = \frac{1}{\omega C_s R_s}$$

(f)

Maxwell bridge

$$L_x = R_a R_b C_s \qquad R_x = \frac{R_a R_b}{R_s}$$

$$Q_x = \omega \frac{L_x}{R_x} = \omega C_s R_s$$

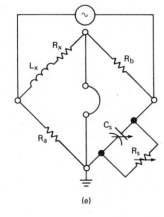

(e)

Owen bridge

$$L_x = C_b R_a R_d$$

$$R_x = \frac{C_b R_a}{C_d} - R_c$$

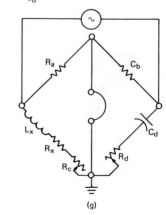

(g)

Figure 6.26 *(Continued)*

161

Substitution method for high impedances

Initial balance (unknown terminals x – x open):

C'_s and R'_s

Final balance (unknown connected to x – x):

C''_s and R''_s

Then when $R_x > 10/\omega C'_s$, there results, with error < 1 percent,

$$C_x = C'_s - C''_s$$

The parallel resistance is

$$R_x = \frac{1}{\omega^2 C'^2_s (R'_s - R''_s)}$$

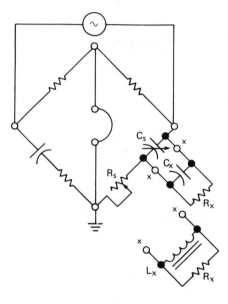

If unknown is an inductor,

$$L_x = -\frac{1}{\omega^2 C_x} = \frac{1}{\omega^2 (C''_s - C'_s)}$$

(h)

Resonance bridge

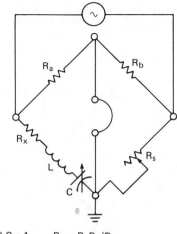

$$\omega^2 LC = 1 \qquad R_x = R_s R_a / R_b$$

(i)

Mutual-Inductance capacitance balance

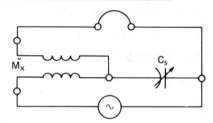

Using low-loss capacitor. At the null $M_x = 1/\omega^2 C_s$

(j)

Felici mutual-inductance balance

At the null:

$M_x = - M_s$

Useful at lower frequencies where capacitive reactances associated with windings are negligibly small.

(k)

Figure 6.26 (*Continued*)

162

Hybrid-coil method

At null: $Z_1 = Z_2$

The transformer secondaries must be accurately matched and balanced to

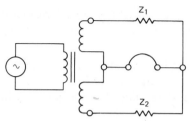

ground. Useful at audio and carrier frequencies.

(l)

Q of resonant circuit by bandwidth

For 3-decibel or half-power points. Source loosely coupled to circuit. Adjust frequency to each side of resonance, noting bandwidth when

$v = 0.71 \times$ (v at resonance)

$$Q = \frac{(\text{resonance frequency})}{(\text{bandwidth})}$$

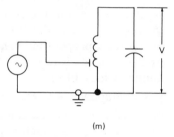

(m)

Q-meter (Boonton Radio type 160A)

R_1 = 0.04 ohm
R_2 = 100 megohms
V = vacuum-tube voltmeter
I = thermal milliammeter
$L_x R_x C_0$ = unknown coil plugged into COIL terminals for measurement.

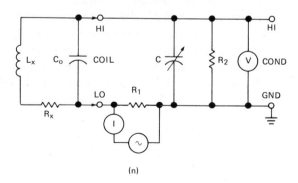

(n)

Figure 6.26 (*Continued*)

163

REVIEW QUESTIONS

1. A 500 V dc voltage is applied across a 1-μF capacitor with a resultant dc current flow of 0.8 μA. Determine the leakage resistance of the capacitor.

2. A capacitor, when tested by a capacitor bridge, was found to have a capacitance of 0.002 μF and a dissipation factor (D) of 0.02.
    a. If the bridge frequency was 1 kHz, determine the resistance R_s in series with the capacitor.
    b. Determine the capacitor power factor (F_p)

3. An inductor has an equivalent circuit of 40-Ω resistance in series with 0.05-H inductance.
    a. Determine the quality factor (Q) of the coil when tested with a 1-kHz bridge.
    b. What is the dissipation factor (D) of the coil?

4. An inductor has a self-resonant frequency of 1 MHz.
    a. At approximately what frequency range should the inductor bridge be set for a proper inductance measurement?
    b. If the inductor is tested with a frequency of 10 MHz, what will the inductor look like?

5. Why is a component's ac resistance higher than its dc resistance?

6. A coil was found to have a complex impedance of $(4 + j6)$ Ω.
    a. What is the quality factor (Q) of the coil?
    b. If the impedance was determined at a frequency of 5 kHz, what is the component's inductance?

7. A 10-μH coil with a self-resonating frequency of 300 MHz is to be measured. What type of instrument should be used to determine the coil characteristics at a frequency of 20 MHz?

8. A series resonant circuit consists of a resistor, capacitor, and inductor. Prove that the voltage amplitude across the inductor is equal to QE, where E is the amplitude of the input voltage.

9. A 10-μH coil is to operate in a circuit where the frequency will be 10 MHz. The coil characteristics (Q, R, L) are measured with a 1-kHz inductance bridge. Will the coil have the measured characteristics at 10 MHz? Explain what will happen to the coil at 10 MHz in terms of R, L, and Q.

7

BASIC SEMICONDUCTOR MEASUREMENTS

7.1 INTRODUCTION

Semiconductor devices such as diodes, thyristors, and transistors form a specialized class of electronic components which require special test equipment and circuits in order to perform valid and meaningful measurements. Initially in this chapter we shall discuss measurements involving devices such as rectifiers and Zener diodes, and then we shall proceed to the characteristics of the silicon control rectifier thyristor device, which is used in power control systems. Finally, we shall develop measurement techniques for amplifying devices such as the bipolar transistor and the field-effect transistor.

7.2 DIODE (RECTIFIER) MEASUREMENTS

The diode is a two-terminal device which exhibits a high resistance when it is reverse biased (the anode voltage is negative with respect to the cathode) and a low resistance in series with a low voltage drop when the device is forward biased. See Fig. 7.1. If the device is made of germanium, its forward voltage (V_f) will be approximately 0.3 V, and if it is of a silicon material, 0.7 V.

7.2.1 Diode Ohmmeter Test

A diode exhibits a characteristic of a high or low resistance depending on how it is biased. A simple test to determine if the diode is acting properly

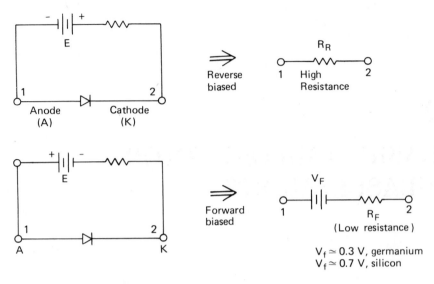

Figure 7.1 Equivalent circuit of a diode under reverse- and forward-biased conditions.

(to determine if the diode is shorted or open) is the diode ohmmeter test. See Fig. 7.2. With the positive probe of the ohmmeter connected to the anode and the negative to the cathode, the ohmmeter should read a resistance from zero to several hundred ohms depending on the diode's characteristics. With the ohmmeter probes reversed the resistance should be 10 to 1,000 times larger. A diode that is shorted will measure low resistance in both directions, and an open diode will measure a very high resistance in both directions.

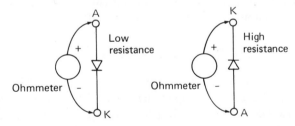

Figure 7.2 Simple short/open circuit test with an ohmmeter.

7.2.2 Diode Static Characteristics

A diode specifically utilized to convert (rectify) ac into a dc voltage (called a rectifier) must meet two important criteria in order to be acceptable. With the device forward biased and a specified current passing through the device, the forward voltage drop (V_f) across the diode should not exceed a specified maximum. With the diode reversed biased by a specified reverse bias voltage, the leakage current flowing must be below a maximum limit, usually specified by the manufacturer.

A test circuit to measure the forward voltage drop (V_f) characteristic is shown in Fig. 7.3. The power supply voltage is varied until the specified

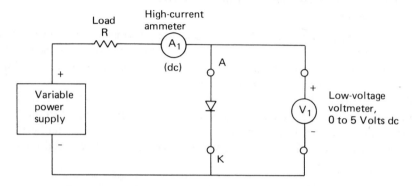

Figure 7.3   Measurement for determining forward voltage drop of a diode.

current as monitored by the ammeter flows through the diode. The voltage drop across the diode (V_1) should be below a specified maximum voltage. The ratio of voltage (V_1) to current (A_1) also determines the forward resistance (R_f) of the diode.

The reverse leakage current may be measured with the circuit shown in Fig. 7.4. Referring to Fig. 7.4, the power supply voltage is adjusted to a specified reverse voltage, and the leakage current is noted with the micro- or milliammeter (A). The measured current should be below a specified maximum for the particular diode. The ratio of the reverse voltage (V) to

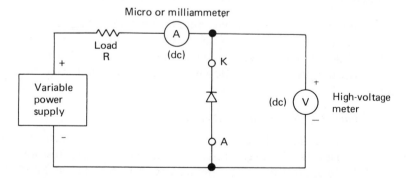

Figure 7.4   Measurement for determining reverse leakage current of a diode.

leakage current determines the reverse resistance (R_r) of the diode. Note that in this measurement the current measuring device may have to be a micro-ammeter, while the voltmeter should have the capability of measuring a high voltage.

7.2.3 Dynamic Measurements of Power Diodes (Rectifier)

The measurements previously discussed provide information on how a diode operates under static or dc conditions. A more realistic test is to measure the diode characteristics under alternating current, or pulsed conditions. This measurement is referred to as dynamic measurements.

Power diodes which are used for rectification of ac into dc may be dynamically tested as follows. An ac voltage whose value is equal to the peak inverse voltage of the diode is applied across the diode with a small resistor in series with it. See Fig. 7.5. If the voltage across the diode is fed into the

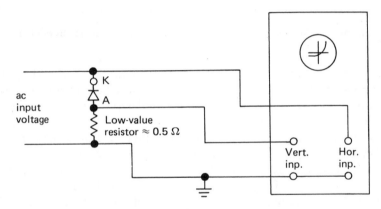

Figure 7.5 Circuit for determining dynamic characteristics of a diode.

horizontal input of the oscilloscope while the resistor voltage is monitored by being fed into the vertical input, a voltage-current characteristic as depicted in Fig. 7.6 is displayed on the oscilloscope. Note that the resistor voltage (vertical input to the oscilloscope) is actually monitoring the diode current.

7.2.4 Pulse Characteristics of Diodes

Diodes are also extensively used in switching circuits in which the input signals are pulses. Diodes may be used as clippers, clampers, restorers, etc. A typical diode application as a clipping circuit is depicted in Fig. 7.7(a). The input signal, which consists of positive and negative pulses when fed into the diode clipper circuit, will be clipped, resulting in an output [Fig.

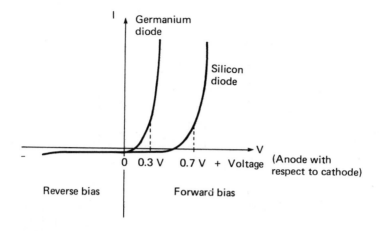

Figure 7.6 Typical volt-ampere characteristic of a germanium and silicon diode.

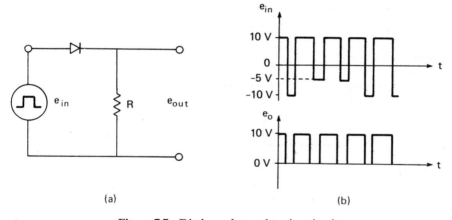

Figure 7.7   Diode used as a clamping circuit.

7.7(b)] of positive pulses. Usually diodes utilized for pulse applications are known as small-signal diodes. In addition to the static test as performed on rectifiers, an important test is to determine the diode recovery time. When the diode is reverse biased by a pulse, there is a time delay before the reverse current settles to its steady-state value. A diode with a long recovery time will be useless for circuits in which the input pulses are operating at a high repetition rate or narrow pulse width. If the diode recovery time is larger than the pulse width, the pulse will be greatly distorted. A circuit which can be used to measure the diode recovery time is given in Fig. 7.8(a), where R_b is adjusted such that the diode is conducting at its proper forward current. When the signal (pulse) input goes negative, the diode will be reverse-biased

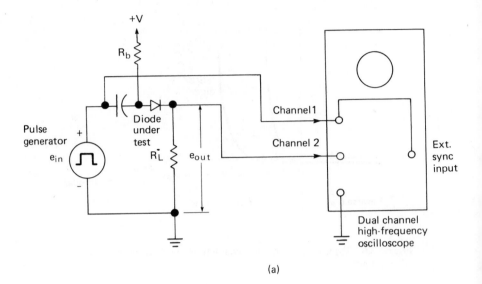

(a)

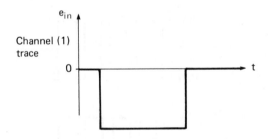

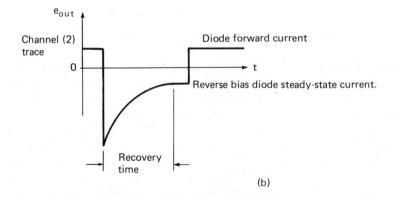

Recovery
time

(b)

Figure 7.8 (a) Circuit for measuring the recovery time of a diode. (b) Recovery time of diode wave form.

by the pulse, and the time it takes for the current through the diode to equal
its steady-state reverse current is the recovery time of the diode. See Fig.
7.8(b). The oscilloscope should be set for external sync., with the sweep on
alternate, in order to view both signals simultaneously.

7.3   MEASUREMENT OF ZENER DIODES

If the reverse voltage across a diode is continually increased, a point is
reached where the diode will break down, with a large current flowing in the
reverse direction. The point at which the diode breaks down is called the
Zener voltage. The voltage will be constant in this mode, and Zener diodes
are specifically designed to operate in this mode. See Fig. 7.9(a) for a typical
volt-ampere characteristic of a Zener diode. If the Zener is biased in its
forward direction, it behaves as an ordinary rectifier diode. In its reverse

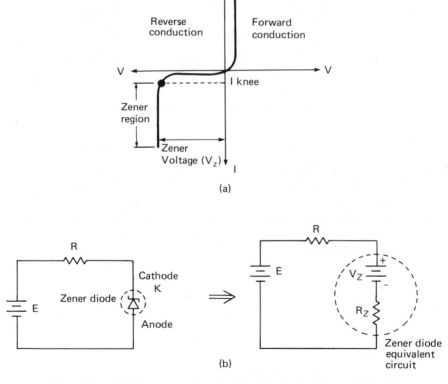

Figure 7.9   (a) Volt-ampere characteristics of a typical Zener diode. (b)
Electron symbol and equivalent circuit of a Zener diode.

direction the maximum current is limited by the power rating of the diode. For proper operation as a voltage regulator the diode must be reversed-biased with a minimum dc current flowing, called I knee. See Fig. 7.9(b) for the electron symbol of a Zener diode and its equivalent circuit. The Zener may be simulated by a battery equal to V_z and a low resistance R_z, which is referred to as the Zener diode dynamic resistance. The smaller R_z, the better the diode will act as a voltage regulator and ac filter. Zener diodes are available with Zener voltages of from 3 to 250 V, with a power rating of from $\frac{1}{4}$ to 50 W.

7.3.1   Measurement of V_z and R_z of a Zener Diode

Important characteristics to determine about a Zener diode are its Zener voltage (V_z) and its dynamic resistance (R_z). A measurement circuit which determines both characteristics is given in Fig. 7.10. The variable power

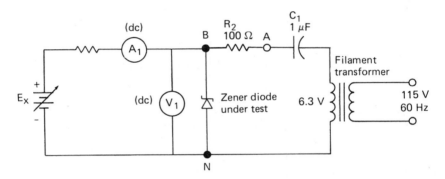

Figure 7.10   Circuit for measuring V_z and R_z of a zener diode.

supply E_x is adjusted until rated dc current (A_1) is flowing through the Zener diode. This current may be 70–80% of its maximum allowable current. The maximum current a Zener can draw is the power rating of the Zener divided by its nominal Zener voltage V_z. The Zener voltage is determined by reading the dc voltmeter V_1. The dynamic resistance R_z is easily determined by measuring the peak-to-peak voltages V_{ab} and V_{bn} with an oscilloscope. Since

$$R_z = \frac{V_{z(\text{ac})}}{I_{z(\text{ac})}}$$

with

$$I_{z(\text{ac})} = \frac{V_{ab}}{R_2}$$

it follows that

$$R_z = \frac{V_{bn}}{V_{ab}/R_2} = \frac{V_{bn}}{V_{ab}/100} = \frac{100V_{bn}}{V_{ab}}$$

Note that R_z will vary as the dc current through the diode is varied.

7.4   SILICON CONTROL RECTIFIER (SCR) MEASUREMENTS—THEORY OF OPERATION

The silicon control rectifier (also referred to as a *PNPN* triode thyristor) is a semiconductor device acting as a controlled switch controlling large amounts of power with a minimum of control power. The SCR is a three-terminal device consisting of an anode, a cathode, and a control gate, as shown in Fig. 7.11. With the SCR forward biased (the anode voltage is positive with

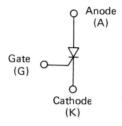

Figure 7.11   Silicon control rectifier (SCR).

respect to the cathode) the device will remain cut off (act as an open switch) until a minimum control gate current (I_{gt}) is applied. With the application of the control gate current, the SCR will conduct (fire) and remain conducting until the anode to cathode voltage is reversed (the SCR becomes back biased) or the current flowing is reduced to a specified minimum, referred to as the holding current (I_h). With the SCR conducting, the control gate loses control of the SCR.

A typical SCR volt-ampere plot is shown in Fig. 7.12. For proper SCR operation the maximum forward bias and reverse bias voltages must be less than the forward blocking and reverse blocking voltages. With this condition and zero gate current the SCR will act as open switch with zero current flowing. With the application of a small amount of gate current (I_{gt}) the SCR will conduct (fire), acting as a closed switch. The level of forward bias voltage (V_{AK}) required for the SCR to conduct is controlled by the amount of gate current. Refer to Fig. 7.12. SCRs are capable of handling currents in excess of 100 A and blocking voltages on the order of 1,000 V. An SCR with an anode current of 100 A can be controlled by a gate current of as little as 50 mA.

7.4.1   SCR Forward Blocking (V_{FBO}) and Reverse Blocking Voltage (V_{RSOM}) Measurements

The forward blocking voltage (V_{FBO}) of an SCR is the voltage that may be applied between the anode and cathode before the device conducts with zero gate current flowing. A circuit which performs this measurement is shown

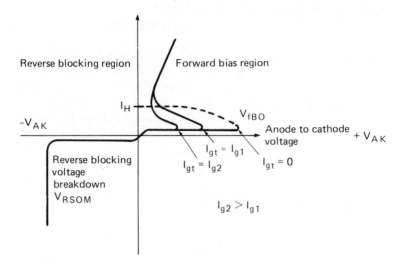

Figure 7.12   SCR volt-ampere characteristics.

in Fig. 7.13. The power supply dc voltage is increased until the SCR fires. When this occurs the anode current (I_{anode}) will suddenly increase to a value limited by the 1,000-Ω resistor. The power supply voltage reading (V_1) when this occurs is the SCR forward blocking voltage.

The reverse blocking voltage (V_{RSOM}) is measured utilizing the circuit of Fig. 7.13 except reversing the SCR anode and cathode connections and utilizing a micro- or millammeter to monitor the current. See Fig. 7.14. The power supply voltage is increased until the anode current increases to a value above a specified maximum. This current may be in the microampere range.

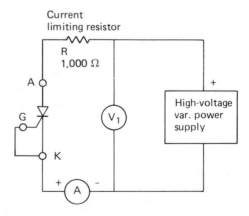

Figure 7.13   Circuit for measurement of SCR forward blocking voltage.

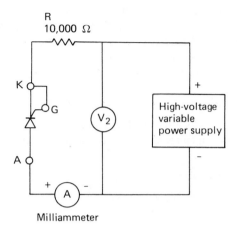

Figure 7.14   Circuit for measurement of SCR reverse blocking voltage.

The power supply reading (V_2) when this occurs is the SCR reverse blocking voltage.

7.4.2   SCR Gate Control Current (I_{gt}) Measurement

An important parameter in determining SCR operation is the gate current I_{gt} required to fire the SCR. The gate current controls the forward bias voltage at which the SCR will fire. Referring to Fig. 7.15 and with the gate voltage (E_g) set to zero and switch S_1 closed, set the power supply (E_{AK}) to a voltage slightly below the SCR forward blocking voltage. The SCR should be cut off. Increase E_g until the SCR fires. The noted gate current (A_1) is the

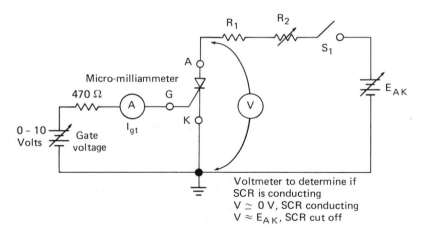

Figure 7.15   Circuit to determine required I_{gt} to fire a SCR.

minimum gate current required to fire the SCR at the particular anode to cathode voltage setting. Open switch S_1 and set the power supply to one-half its previous setting. Close switch S_1 and note that the SCR should be cut off. Increase E_g further until the SCR fires and make note of the new I_{gt}. As the anode to cathode voltage is reduced a larger I_{gt} is required to fire the SCR.

7.4.3 SCR Holding Current (I_h) Measurements

A further important SCR measurement is determining the holding current I_h. This may be easily measured with the circuit of Fig. 7.16. With R_2 set to

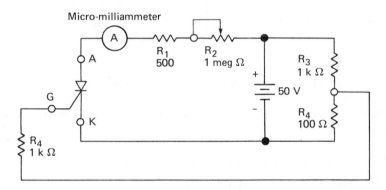

Figure 7.16 Circuit for measuring SCR holding current I_h.

$0\,\Omega$, the SCR will be conducting with a current of approximately 100 mA flowing. The variable resistance R_2 is increased until the SCR cuts off, with the current flow dropping to zero. The current flowing just previous to the SCR cutting off is the holding current (I_h) of the SCR. Depending on the type of SCR, I_h may vary from microamperes to several hundred milliamperes.

7.5 TRANSISTOR MEASUREMENTS

Transistor measurements will consist of two parts: one determining important parameters of a transistor as a device by itself and the other the measurement of transistor circuits such as a common emitter amplifier.

The transistor amplifier operation may be understood if we replace the transistor with its equivalent ac circuit, shown in Figs. 7.17(a) and (b). With the amplifier properly dc-biased, the base to emitter junction consists of a forward-biased diode with the collector to base as a reverse-biased diode. Across the reverse-biased diode a current generator equal to βi_B flows, where β is the forward current gain, which may vary from 10 to 500 depending on the particular transistor. The ac equivalent circuit of a forward-biased diode

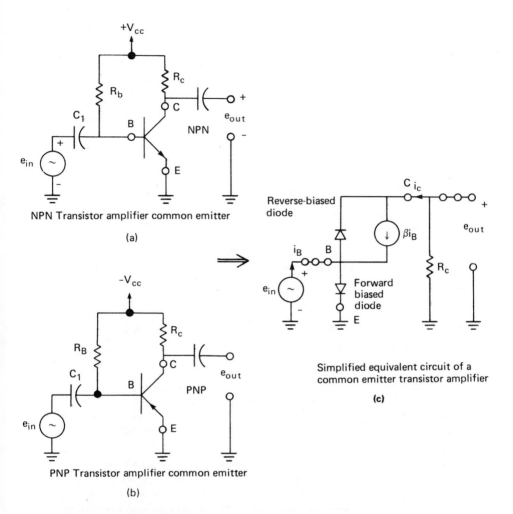

Figure 7.17 (a) NPN transistor amplifier common emitter. (b) PNP transistor amplifier common emitter. (c) Simplified equivalent circuit of a common emitter transistor amplifier.

may be replaced with a small resistor of 10–200 Ω, called r_e, while a back-biased diode can be considered an open circuit. With these approximations the circuit of Fig. 7.18 results. From simple circuit analysis,

$$e_{out} = -\beta i_B R_c \quad \text{and} \quad e_{in} = i_B r_e (\beta + 1)$$

With a typical β on the order of 100, $(\beta + 1) \approx \beta$. Therefore the voltage gain

$$A_v = \frac{e_{out}}{e_{in}} = \frac{-\beta i_B R_c}{i_B r_e \beta} = \frac{-R_c}{r_e}$$

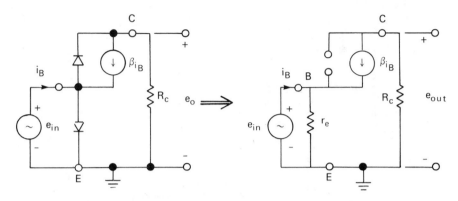

Figure 7.18 Simplified ac equivalent circuit of a common emitter transistor amplifier.

For a typical transistor, R_c is on the order of 2 KΩ; and with r_e being the impedance of a forward-biased diode with a resistance as high as 200 Ω, a voltage gain of $-2,000/200 = -10$ results. The minus sign denotes that the output voltage is 180° out of phase with the input signal.

7.5.1   Measurements of Basic Transistor Characteristics

7.5.1.1   Transistor ohmmeter test.   This test will be a quick measurement to determine if the transistor is open or shorted. This test will not determine any of its particular characteristics.

A transistor may be easily checked out by performing a resistance check from base to emitter, base to collector, and collector to emitter junctions. The base to emitter and base to collector junctions consist of diodes. The ohmmeter reading should consist of a diode action (high with one polarity and low with the reverse polarity) between the base to emitter and base to collector junctions. With the ohmmeter across the collector to emitter junction the meter should read a high resistance in both directions. See Fig. 7.19.

7.5.1.2   Transistor current gain (H_{fe} or β) and current leakage measurement. How a transistor will operate as a linear amplifier is greatly determined by the forward transfer current ratio (in a common emitter configuration), called H_{fe} or beta (β). The β of a transistor will vary widely between transistors of the same type. In general

$$H_{fe} = \beta_{dc} = \frac{I_{c(dc)}}{I_{B(dc)}}$$

We may also assume with little error that the samll-signal ac common emitter current transfer ratio (also referred to as the current gain ratio) $\beta_{ac} = h_{fe}$ is equal to the dc current current gain:

$$H_{fe} = \beta_{dc} = h_{fe} = \beta_{ac} = \beta$$

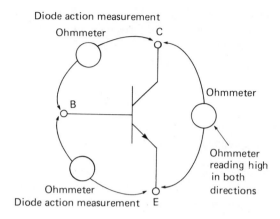

Figure 7.19 Ohmmeter transistor measurement.

A circuit which performs this measurement is depicted in Fig. 7.20. As V_{BB} is increased, read meters $A_1(I_B)$ and $A_2(I_C)$. The transistor β is $A_2/A_1 = I_C/I_B$. For a power transistor, A_2 will be an ammeter, with A_1 a milliammeter. The leakage current of most modern transistors is so minute (on order of 10 nA to 5 μA) that this measurement will be deleted.

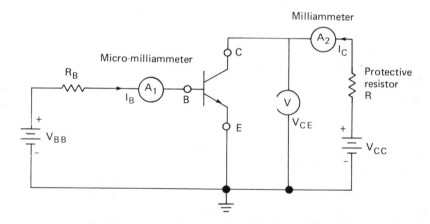

Figure 7.20 Circuit for determining β of a transistor.

7.5.1.3 Measurement of the gain-bandwidth factor f_t.

The previous statement that a transistor current gain A_i is approximately equal to $\beta_{ac} = \beta_{dc}$ is applicable only for transistors operating at a signal frequency much smaller than the gain-bandwidth factor, called f_t. As the ac signal frequency is increased, A_i is initially constant (and equal to β_{dc}) until a frequency at which the A_i falls with frequency. The frequency at which A_i drops to unity is called the gain-bandwidth factor f_t. See Fig. 7.21. With the transistor collector (ac) short-circuited to its emitter, its current gain A_i is found by

$$A_i = \frac{\beta_{dc}}{[1 + (f/f_\beta)^2]^{1/2}}$$

where f = operating signal frequency,
  f_β = signal frequency at which $\beta = 0.707\beta_{dc}$.

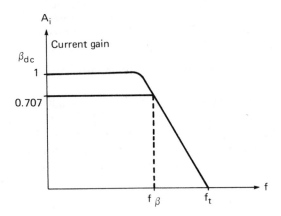

Figure 7.21   Plot of a transistor current gain A_i as a function of frequency.

Operating the transistor at a frequency $f \ll f_\beta$, $A_i = \beta_{dc}$. For the case where $f/f_\beta \gg 1$,

$$A_i = \frac{\beta_{dc}}{f/f_\beta} = \frac{f_\beta \beta_{dc}}{f} = \frac{f_t}{f}$$

At the frequency $f = f_t$, the transistor will be inoperative since at this frequency its $A_i = \beta$ is equal to unity.

Allowing for the current gain A_i to drop to 0.707 of its low-frequency current gain, the maximum signal frequency at which the transistor is to be operated is f_β.

With the gain-bandwidth factor f_t being as high as several hundred megahertz, it will be indirectly measured by determining the frequency (f_β) for which the transistor current gain is 0.707 of its low-frequency (dc) current gain. At this frequency $f_t = (f_\beta)(A_{i(dc)}) = f_\beta \beta_{dc}$.

Utilizing the circuit of Fig. 7.22, V_{BB} is adjusted for a dc collecter current at which f_t is to be measured. Set the generator frequency to 500 Hz and measure e_{AG}, e_{BG}, and e_{CG} with a high-frequency scope. The low-frequency

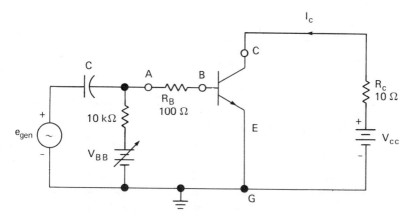

Figure 7.22   Circuit diagram for measuring the gain bandwidth factor (f_t) of a transistor.

current gain is

$$A_i = \beta_{dc} = \beta = \frac{i_c}{i_B} = \frac{e_{CG}/10}{(e_{AG} - e_{BG})/100} = \frac{10e_{CG}}{e_{AG} - e_{BG}}$$

With e_{AG} held constant, increase the frequency until e_{CG} drops to 0.707 of its value at 500 Hz. This frequency is f_β. The gain-bandwidth factor f_t is found by $f_t = \beta_{dc} f_\beta = A_i f_\beta$.

7.5.2   Transistor Amplifier (Small-signal ac) Measurements

7.5.2.1   Common emitter configuration.   The important measurements to be made on a transistor amplifier are the input resistance (R_{in}), the voltage gain (A_v), the output resistance (R_o), and the (3-dB frequency bandwidth.

These measurements are all performed with an oscilloscope on the actual operating amplifier. See Fig. 7.23. Set the frequency of the generator (e_{gen}) to a midband frequency, with its ac signal of such amplitude that the output voltage (e_o) is an undistorted sine wave. Measure the input voltage (e_{in}) with an oscilloscope. Connect a variable resistor R_v between e_{gen} and the transistor input. Set the value of R_v such that e_{in} is one half of the previous e_{in}. Disconnect the variable resistor (R_v) and measure its resistance with an ohmmeter. The measured resistance is the R_{in} of the transistor amplifier. With e_{gen} feeding directly into the amplifier, (R_v deleted), measure e_{out} and e_{in}. The transistor voltage gain A_v is e_{out}/e_{in}. To determine the output resistance R_o, connect a variable resistor (R'_L) across the output and decrease its value until the output voltage (e_{out}) is reduced to one half of its original value. Disconnect R'_L and measure its resistance with an ohmmeter. The measured

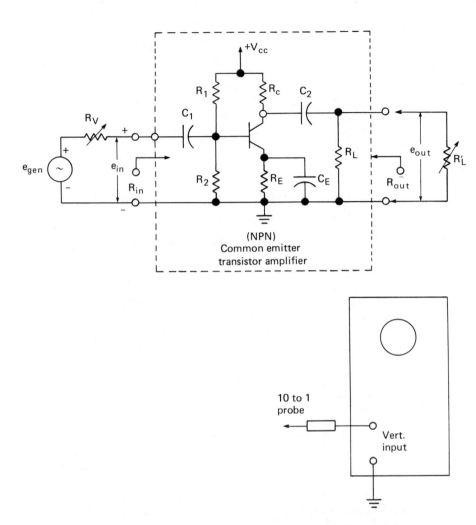

Figure 7.23   Measurements of transistor amplifier characteristics.

resistance is the output resistance (R_o) of the amplifier. The amplifier bandwidth is determined as follows. With the input voltage e_{gen} kept at its initial setting and at a constant amplitude, decrease the generator frequency until the output voltage (e_{out}) drops to 0.707 of its original reading. Note the frequency when this occurs. This frequency is referred to as the low-frequency (f_{low}) limit of the amplifier. Increase frequency until a value is reached where the output again drops off to 0.707 of its original reading. This frequency is referred to as the high-frequency limit (f_{high}) of the amplifier. The amplifier bandwidth is $f_{high} - f_{low}$. See Fig. 7.24.

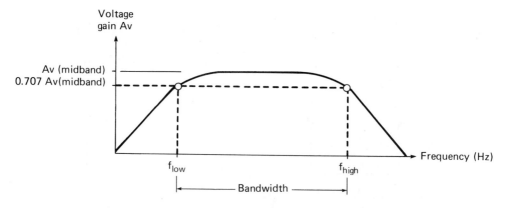

Figure 7.24   Typical frequency response of a transistor amplifier.

7.6   FIELD-EFFECT TRANSISTOR

The field-effect transistor is a semiconductor device which controls the flow of current by application of an electric field rather than a direct biasing voltage. There are two basic types of field-effect transistors: the *junction field-effect transistor* (JFET) and the *metal-oxide semiconductor field-effect transistor* (MOSFET). The FET characteristics differ from the conventional transistor in that its input resistance is much higher, usually in the region of hundreds of megohms. The main disadvantage of the FET is a lower gain and bandwidth response as compared with a conventional transistor.

7.6.1   JFET—Theory of Operation

Similar to conventional transistors which are *NPN* or *PNP*, a JFET may be an *N*-channel or a *P*-channel device. The three terminals of the JFET are the gate, drain, and source. See Fig. 7.25 for a comparison between conventional transistors and JFETs with respect to proper biasing voltages.

To analyze the JFET as an ac amplifier we convert the JFET to its equivalent circuit, as shown in Fig. 7.26. For a typical JFET r_d is the internal resistance of a JFET and may vary from 500 to 20 KΩ depending on the device. The transconductance g_m is a function of the dc bias current I_D flowing and ranges from 500 to 10,000 μmhos. From the ac equivalent circuit of the JFET (Fig. 7.26), $e_o = -g_m e_{gs} R'_D$, where

$$R'_D = \frac{r_d R_D}{r_d + R_D}$$

Since e_{gs} is equal to e_{in},

$$e_o = -g_m e_{in} R_D \quad \text{or} \quad A_v = \frac{e_{out}}{e_{in}} = -g_m R'_D$$

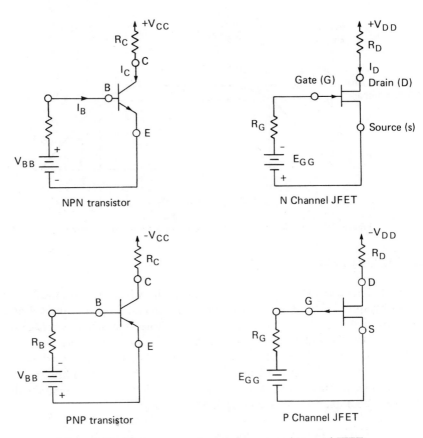

Figure 7.25 Biasing comparisons between transistor and JFETs.

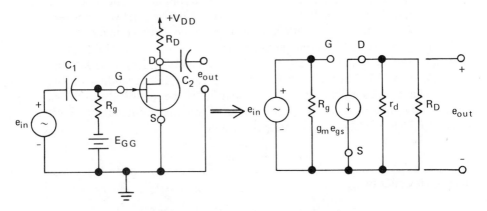

Figure 7.26 Equivalent ac circuit of a JFET.

184

The minus sign denotes that the output signal is 180° out of phase with the input signal.

7.6.2 Measurement of JFET Characteristics

The important characteristics to determine for a JFET are I_{DSS}, V_p, and g_m. For a JFET common source amplifier, important measurements are its voltage gain A_v, output resistance R_o, and 3-dB bandwidth. The JFET input resistance is on the order of hundreds of megohms ($10^{11}\ \Omega$) and may be assumed to be equal to infinity.

7.6.2.1 Measurement of I_{DSS} of a JFET.   The saturation drain current I_{DSS} is the maximum allowable dc drain current and is the drain current that flows with $V_{GS} = 0$ V. This current is a function of V_{DS} when V_{DS} is less than the pinch-off voltage V_p. If V_{DS} is equal to or greater than V_p, then I_{DSS} becomes independent of V_{DS} and varies with the gate to source voltage V_{GS}. Refer to Fig. 7.27. The pinch-off voltage V_p is the gate to source voltage V_{GS} for I_D equal to zero.

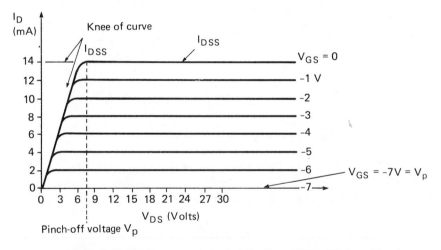

Figure 7.27   Volt ampere characteristics for a JFET, (*N* channel).

To measure the I_{DSS} of a JFET, wire the circuit of Fig. 7.28. With the gate to source shorted together ($V_{GS} = 0$) the current (A_1) that flows in the drain is I_{DSS}. E_{DD} is set such that V_{DS}, the typical operating voltage of the JFET, is greater than the pinch-off voltage (V_p) and less than any of the breakdown voltages of the JFET.

7.6.2.2 Determining the pinch-off voltage V_p of a JFET.   The pinch-off voltage V_p may be measured by the circuit of Fig. 7.29. E_{GG} is increased until $I_D = 0$. The voltage amplitude of E_{GG} (V_1 reading) required to make $I_D = 0$ is the pinch-off voltage V_p of the JFET.

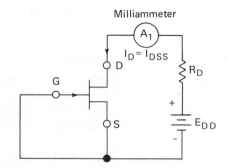

Figure 7.28   Circuit to measure I_{DSS} of a JFET.

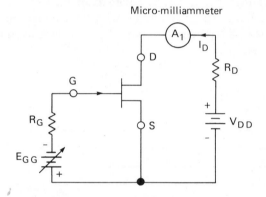

Figure 7.29   Circuit to measure V_p of a JFET.

7.6.2.3   Measurement of JFET transconductance (g_m).

The transconductance of a JFET is an expression which relates the change in output current which may be induced by a change in the input voltage (V_{GS}),

$$g_m = \frac{\Delta I_D}{\Delta V_{GS}}$$

at a given drain to source voltage. The g_m of a JFET is typically higher in the region where the drain current I_D is high. For V_{DS} greater than V_p, the g_m is independent of the drain to source voltage. The g_m measurement is determined by wiring the circuit of Fig. 7.30. Adjust E_{GG} and R_D such that V_{DS} is greater than the pinch-off voltage V_p with a typical drain current I_D flowing. Vary E_{GG} slightly above and below its previous setting, noting the change in drain current (A_1) as a function of E_{GG}. Monitor the E_{GG} change with a voltmeter (V_1) connected from gate to source. The g_m of the JFET is determined by

$$g_m = \frac{\Delta I_{DS}}{\Delta V_{GS}} = \frac{\Delta I_{DS}}{\Delta E_{GG}} = \frac{\text{change in } A_1}{\text{change in } V_1}$$

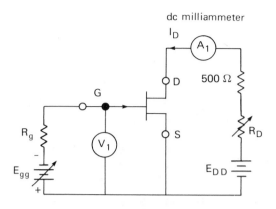

Figure 7.30   Circuit to measure g_m of a JFET.

7.6.2.4   Measurement of JFET common source amplifier characteristics.
The important measurements to be made on a JFET amplifier are its voltage
gain (A_v), output resistance (R_o), and 3-dB frequency bandwidth. These
measurements are made exactly as described in Section 7.5.2.1. Wire the
amplifier circuit of Fig. 7.31. With the input signal (e_{gen}) set to the midband
frequency and with an amplitude such that the output (monitored with an
oscilloscope) is undistorted, measure e_{out} and e_{in}. The voltage gain $A_v =
e_{\text{out}}/e_{\text{in}}$. The output resistance R_o is determined by shunting the output with
a variable resistor (R_L) with such a value that e_o is one-half the voltage
measured previously. The resistance value of R_L is equal to the R_o of the
amplifier. The bandwidth is determined as follows: With the input voltage
held constant, decrease the generator frequency until the output voltage
decreases to 0.707 of its original value. Note the frequency, which is the flow
of the amplifier. Proceed to increase the frequency until e_{out} decreases to
0.707 of its midband voltage. Note this new frequency, which is f_{high} of the
amplifier. The bandwidth is $f_{\text{high}} - f_{\text{low}}$.

7.6.3   MOSFET Theory of Operation and Measurements

The metal-oxide semiconductor field-effect transistor (MOSFET), which
also contains a drain, source, and gate, acts similarly to the JFET except as
follows: In a MOSFET a thin layer of insulating material (silicon dioxide)
is placed over the channel before deposition of the gate electrode. This makes
the MOSFET input impedance much higher than that of a JFET. Typical
input impedance of a MOSFET is on the order of 10^{14} Ω. See Fig. 7.32 for
circuit diagrams of different types of MOSFETs. A MOSFET may operate
in the depletion or enhancement mode, while a JFET operates only in its
depletion mode. See Fig. 7.33 for typical volt-ampere characteristics and
I_D vs. V_{GS} curves of MOSFETs.

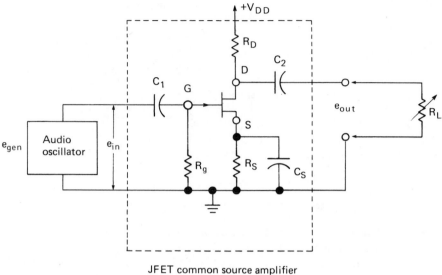

JFET common source amplifier

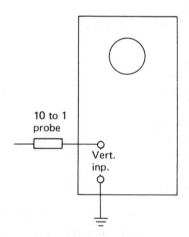

Figure 7.31   Circuit to measure JFET amplifier characteristics.

A depletion mode MOSFET is a device which is normally on, since with a zero gate bias voltage, a large drain current (I_{DSS}) flows. A MOSFET which operates in the enhancement mode will have zero current flowing with $V_{GS} = 0$ V. With the application of a positive voltage the MOSFET will conduct. The minimum gate to source voltage required for the MOSFET to conduct is called the threshold voltage V_T. MOSFETs are now available with V_T as low as 0.7 V. MOSFETs are also available which can operate in the depletion or enhancement mode. Due to the high gate to source imped-

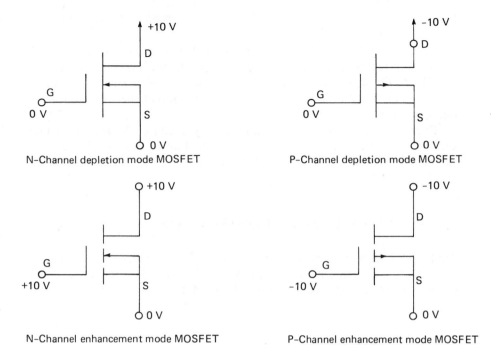

Figure 7.32 MOSFET symbols and typical biasing voltages.

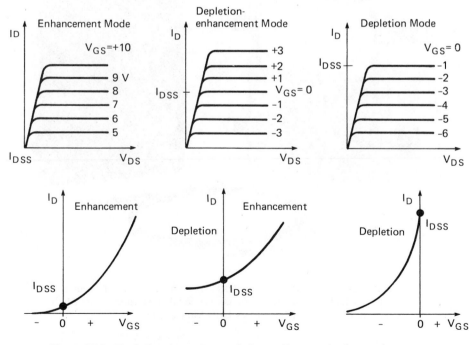

Figure 7.33 Typical volt-ampere and I_D v. V_{GS} curves for various MOSFETs.

ance ($10^{14}\ \Omega$) a MOSFET can be destroyed due to static voltage fields. It is therefore very important how a MOSFET is handled or how we solder a MOSFET. On a dry day a static voltage buildup on a person's hands of greater than 100 V is possible. MOSFETs have found their greatest use in digital circuits.

Except for the different modes and biasing procedure a MOSFET operates similarly to a JFET with the same small-signal ac equivalent circuit. Measurement procedures of the MOSFET parameters are similar to those of the JFET.

7.7   SEMICONDUCTOR TEST INSTRUMENTS

Transistor characteristics such as the beta (β) or leakage current may be tested with a semiconductor tester, as shown in Fig. 7.34. This tester has the capabilities of testing a transistor (or FET) in circuit or out of circuit. For

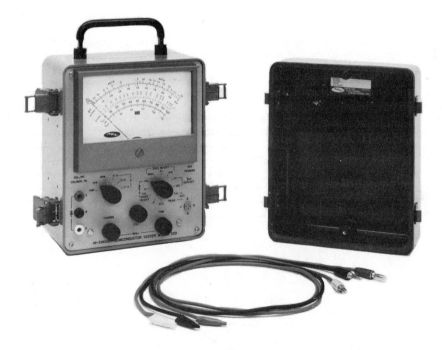

Figure 7.34   In-circuit transistor tester measures both out-of-circuit and in-circuit, and also covers both conventional transistors and FETs. Courtesy of American Electronics Labs Model 259c.

the tester to measure accurately a transistor or FET in circuit, the circuit resistors and components loading the device must be larger than a specified minimum.

An instrument which can test semiconductors under dynamic conditions is a transistor curve tracer oscilloscope display, shown in Fig. 7.35. The instrument is able to present a display of voltage-ampere characteristics of diodes, Zener diodes, SCRs, transistors, and FETs. Steps of current or voltage in staircase form are selected by the corresponding control and applied to the transistor or FET input while the device output voltage is swept, producing a family of volt-ampere curves. The curves are calibrated, enabling us to determine the transistor beta (β) or the tranconductance (g_m) of an FET. See Figs. 7.36(a) and (b) for typical voltages and currents applied to a transistor with their resulting displays.

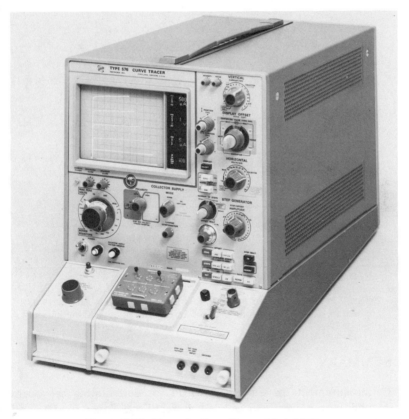

Figure 7.35   Tetronix Curve Tracer Model 576.

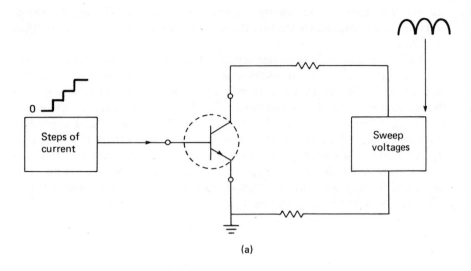

(a)

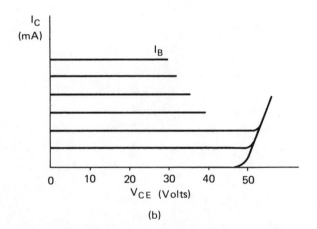

(b)

Figure 7.36 (a) Transitor under test. (b) Typical voltages and currents applied to a transistor from a curve tracer and the resulting output display.

7.8 SUMMARY

Initially measurements have been described for determining the operation of semiconductor power devices such as rectifiers, Zener diodes, and the silicon control rectifier.

Measurements for determining the characteristics of the bipolar junction transistor and field-effect transistor were then introduced.

Transistor amplifiers (BJTs and FETs) were tested, determining the following: voltage gain, input and output impedance, and bandwidth.

The material in this chapter allows the technician or engineer to easily troubleshoot or test basic semiconductor devices utilizing simple laboratory instrumentation.

REVIEW QUESTIONS

1. A reverse voltage of 600 V was applied across a diode, resulting in a current flow of 3.5 μA. What is the diode's reverse resistance?

2. a. What is the voltage across a 10-V, 1-W Zenei diode when it is forward biased?
   b. What is the maximum allowable current through the Zener diode?

3. The following specifications are given for an SCR: $V_{RSOM} = 200$ V, $V_{FBO} = 160$ V, $I_{\text{holding}} = 20$ mA.
   a. With zero gate current, determine the minimum voltage across the SCR (V_{AK}) in order for it to fire.
   b. If the current through the SCR is reduced below 20 mA, what will occur?

4. The following conditions were determined for a transistor when checked with an ohmmeter:
   a. $R_{B\text{-}E}$ = low, $R_{E\text{-}B}$ = high, $R_{B\text{-}C}$ = high, $R_{C\text{-}B}$ = high, $R_{C\text{-}E}$ = high, $R_{E\text{-}C}$ = high. Is the transistor operating properly? Explain your answer.
   b. $R_{B\text{-}E}$ = high, $R_{E\text{-}B}$ = low, $R_{B\text{-}C}$ = low, $R_{C\text{-}B}$ = high, $R_{C\text{-}E}$ = high, $R_{E\text{-}C}$ = low. Is the transistor operating properly? Explain your answer.

5. The following measurements were performed on a transistor amplifier: $I_{B(\text{dc})} = 100$ μA, $I_{C(\text{dc})} = 5$ mA. The transistor output voltage at no load was 4 V peak to peak and dropped to 1 V peak to peak when a 2-kΩ resistor was connected across it.
   a. What is the low frequency beta (β_{dc}) of the transistor?
   b. Determine the R_o of the transistor amplifier.

6. The gain of an amplifier dropped 3 dB. If the initial gain was 60, what is the final gain?

7. The following tests were made on a transistor amplifier: $A_{v(\text{midband})} = 200$, $f_{\text{low}(3\text{dB})} = 100$ Hz, $f_{\text{high}(3\text{dB})} = 30$ kHz.
   a. What is the amplifier gain at an input frequency of 5 kHz?

  b. Determine the amplifier bandwidth.
  c. What is the amplifier gain at a frequency of 30 kHz?

8. The following measurements were performed on an N-channel JFET
    amplifier: (1) $V_{GS} = -5$, $I_D = 0$ mA; (2) $V_{GS} = 0$, $I_D = 7$ mA; (3)
    when V_{GS} was changed from -4 to -3 V, the drain current increased
    10 mA. Determine the following: (a) the pinch-off voltage (V_p), (b)
    the saturation drain current (I_{DSS}), and (c) the transconductance (g_m)
    of the JFET.

9. The input voltage to an N-channel JFET amplifier was 300 mV peak
    to peak when connected directly to a signal generator. When a resistor
    of 3 mΩ was placed in between the generator and JFET the input volt-
    age to the amplifier dropped to 150 mV. What is the amplifier input
    impedance?

10. What precautions must be taken when performing measurements on
    a MOSFET?

8

INTEGRATED CIRCUIT
MEASUREMENTS

8.1 INTRODUCTION

Electronic circuits can be broadly divided into two basic areas, analog and digital.

In digital circuits, the active elements are used primarily as switches, being operated in the "on" or "off" state but not in between. In analog circuits, various input signals will produce a corresponding output in which there is frequently a linear relationship between the input and output. Integrated circuits (abbreviated IC's) operating in this mode are called linear integrated circuits. Note that many linear IC's may be operated in a nonlinear mode.

In this chapter we shall describe test measurements of two basic IC blocks which form an integral part of many different and complex ICs. For linear ICs, characteristics of the *operational amplifier* will be determined, while for digital ICs, measurements of a basic transistor-transistor logic (TTL) gate will be demonstrated.

A student who thoroughly understands the operation and measurement techniques of these two basic IC devices will be capable of comprehending the literally hundreds of different linear and digital ICs available.

8.2 OPERATIONAL AMPLIFIER—THEORY OF OPERATION

The term *operational amplifier* (abbreviated op amp) was originally coined for those amplifiers of an analog computer performing mathematical operations such as integration, summation, and subtraction. Today, the widest use of op amps is in such applications as signal conditioning, servo and process controls, analog instrumentation, impedance transformation, active filters, and the like.

The great versatility and many advantages of op amps stem from the use of negative feedback. Negative feedback tends to improve the amplifier's gain stability, reduce the output impedance, and increase the input impedance and bandwidth.

8.2.1 Ideal Operational Amplifier

In many ways an IC op amp approaches an ideal amplifier which has the following characteristics:

1. Input resistance $R_{in} = \infty$.
2. Output resistance $R_o = 0$.
3. Voltage gain $A_{VOL} = -\infty$.
4. Bandwidth $= \infty$.
5. $V_o = 0$ when $(V_-$ input$) - (V_+$ input$) = 0$.

The electronic circuit symbol of a basic ideal op amp is shown in Fig. 8.1(a).

With an infinite open loop voltage gain (A_{VOL}), the difference voltage between the two inputs must be 0 V. The inputs therefore are said to have a virtual short circuit between them. The current through the *virtual short* is also zero, due to the infinite input resistance of the op amp. For purposes of analysis, the inverting amplifier [Fig. 8.1(a)] may be replaced with the equivalent circuit shown in Fig. 8.1(b).

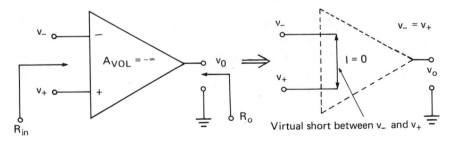

Figure 8.1 (a) Basic op-amp circuit. (b) Equivalent circuit of an op amp.

Shown in Fig. 8.2(a) and (b) is an ideal op amp converted into an inverting amplifier by utilizing negative feedback.

Writing Kirchhoff's current law at the $(-)$ input [Fig. 8.2(b)],

$$I_1 + I_f = 0$$

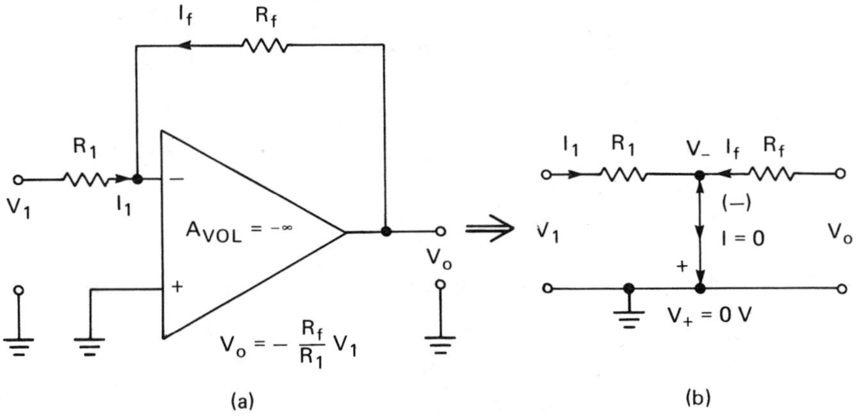

Figure 8.2   (a) Inverting amplifier. (b) Equivalent circuit of an inverting amplifier.

With a virtual short between $-$ and $+$ inputs,

$$V_- = V_+ = 0$$

$$I_1 = \frac{V_1 - V_-}{R_1} = \frac{V_1 - 0}{R_1}$$

$$I_f = \frac{V_o - V_-}{R_f} = \frac{V_o - 0}{R_f}$$

Substituting,

$$\frac{V_1}{R_1} + \frac{V_o}{R_f} = 0$$

$$\frac{V_o}{V_1} = A_v = -\frac{R_f}{R_1}$$

Note that the gain of the inverting amplifier circuits is independent of the open loop gain but a function of two resistors R_f and R_1.

An operational amplifier with proper feedback may also be converted to a noninverting amplifier. See Fig. 8.3(a). With a virtual short between V_+ and V_-, the circuit of Fig. 8.3(a) can be converted to the circuit shown in Fig. 8.3(b). Writing Kirchhoff's current law at the V_- inputs [Fig. 8.3(b)],

$$-I_1 + I_f = 0$$

$$I_1 = I_f$$

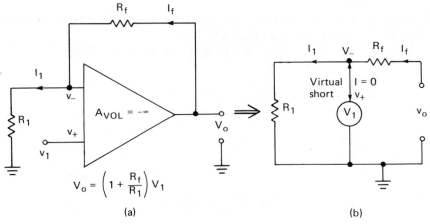

Figure 8.3   (a) Non-inverting amplifier. (b) Equivalent circuit of a non-inverting amplifier.

with $V_- = V_+ = V_1$,

$$I_1 = \frac{V_-}{R_1} = \frac{V_1}{R_1}$$

$$I_f = \frac{V_o - V_-}{R_f} = \frac{V_o - V_1}{R_f}$$

Substituting,

$$\frac{V_1}{R_1} = \frac{V_o - V_1}{R_f}$$

$$\frac{V_o}{V_1} = A_v = 1 + \frac{R_f}{R_1}$$

Similar to the inverting amplifier, the gain for a noninverting amplifier is a function of the feedback resistors and is independent of the open loop gain (A_{VOL}).

8.2.2   Nonideal Operational Amplifier

The characteristics of an actual op amp differ from an ideal op amp as follows:

1. Input resistance $R_i = 10^4 – 10^9 \ \Omega$.
2. Output resistance $R_o = 10 – 5 \ k\Omega$.
3. Voltage gain (A_{VOL}) $= 10^4$ to 10^9.
4. Limited bandwidth.

With a finite open loop voltage gain (A_{VOL}), the gain of an inverting op amp circuit with feedback is given by

$$A_v = -\frac{R_f}{R_1}\left\{\frac{1}{1 + (1/A_{VOL})[(R_1 + R_f)/R_1]}\right\}$$

The term $R_1/(R_1 + R_f)$ is the feedback factor, which is the ratio of output voltage fed back to the input, and is called β:

$$\beta = \frac{R_1}{R_1 + R_f}$$

We may rewrite the first equation as

$$A_v = -\frac{R_f}{R_1}\left[\frac{1}{1 + (1/A_{\text{VOL}}\beta)}\right]$$

The term $A_{\text{VOL}}\beta$ is referred to as the loop gain and is extensively used in determining how a nonideal op amp operates:

$$\text{loop gain} = A_{\text{VOL}}\beta$$

If the loop gain ($A_{\text{VOL}}\beta$) is much greater than 1, the voltage gain for an inverting amplifier reduces to $-(R_f/R_1)$, which is the gain for an ideal inverting op amp. While feedback reduces the op amp gain by the loop gain ($A_{\text{VOL}}\beta$) factor, there is a corresponding increase in all other amplifier specifications. An amplifier's stability, input resistance, bandwidth, and output resistance are all improved as the loop gain increases:

$$R_{\text{in}f} = R_{\text{in}}(1 + A_{\text{VOL}}\beta)$$

$$\text{BW}_f = \text{BW}(1 + A_{\text{VOL}}\beta)$$

$$R_{of} = \frac{R_o}{1 + A_{\text{VOL}}\beta}$$

where $R_{\text{in}f}$, BW_f, and R_{of} are the input resistance, bandwidth, and output resistance with feedback.

8.3   OPERATIONAL AMPLIFIER CHARACTERISTICS AND MEASUREMENTS

The important characteristics of a typical operational amplifier are listed in Table 8.1. Defenitions, test procedures, and circuits will be developed for determining these characteristics. It will be assumed that all op-amp devices are initially dc biased for proper operation.

8.3.1   Input Offset Voltage (V_{OS}), Offset Current (I_{OS}), and Bias Current (I_B)

When the op amp is utilized as a dc amplifier, it is important that for a zero dc input voltage the output is 0 V. In actuality, there will be an output dc voltage (called the offset voltage) for a zero input voltage. This offset voltage, which is a serious limitation in the accuracy of a dc amplifier, is caused by a mismatch at the differential amplifier input stage.

The three parameters which cause the output offset voltage as specified

Table 8.1

Typical Parameters of an IC Operational Amplifier

Input offset voltage	V_{OS}	2 mV
Input offset current	I_{OS}	10 nA
Input bias current	I_B	100 nA
Input offset voltage temperature drift	V_{OS}	5 μV/°C
Input offset current temperature drift	I_{OS}	10 nA/°C
Open loop dc voltage gain	A_{VOL}	100,000
Open loop 3-dB bandwidth	BW_{OL}	5 kHz
Unity gain bandwidth	f_t	0.5 mHz
Slew rate	S_r	I V/μs
Differential input impedance	R_{in}	100 KΩ
Common mode input impedance	R_c	10 MΩ
Output impedance	R_o	50 Ω
Common mode rejection ratio	CMRR	80 dB

by op amp manufacturers are the following: input offset voltage (V_{OS}), input offset current (I_{OS}), and input bias current (I_B).

Input offset voltage (V_{OS}).   A mismatch in the base to emitter voltage of the input differential amplifier transistors is the cause of a dc offset voltage at the output.

Manufacturers specify the input offset voltage (V_{OS}) as the required input to the op amp such that the output offset voltage is zero.

When an op amp is utilized as an inverting or noninverting amplifier it will have a dc output offset voltage due to V_{OS}, given by

$$V_{o(offset)} = V_{os}\left(1 + \frac{R_f}{R_1}\right)$$

With $1 + (R_f/R_1) = (R_1 + R_f)/R_1 = 1/\beta$,

$$V_{o(offset)} = \frac{V_{os}}{\beta}$$

Input bias (I_B) and offset current (I_{OS}).   For an op amp to operate properly, it is necessary to supply a dc bias current (pA to μA) at the input differential amplifier stage. The average of the two bias currents is referred to as the input bias current (I_B). With reference to Fig. 8.4,

$$\text{bias currrent} = I_B = \frac{I_{B1} + I_{B2}}{2}$$

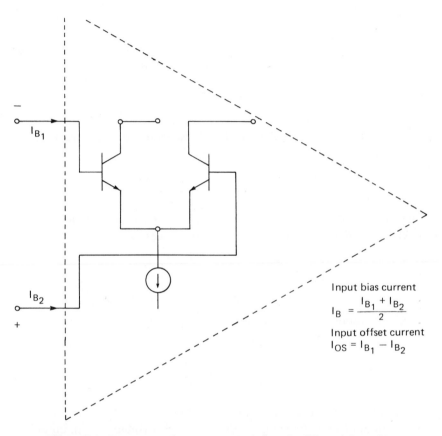

Figure 8.4   Input stage of an op amp consisting of a differential amplifier.

The two input transistors are usually not perfectly matched, and a different bias current will be required at each input. The difference between the two input bias currents is referred to as the input offset current (I_{OS}):

$$\text{input offset current} = I_{OS} = I_{B1} - I_{B2}$$

With the op amp operating as an inverting or noninverting amplifier [refer to Figs. 8.5(a) and (b)] there will be an output dc offset voltage due to the input bias currents given by

$$V_{o(\text{offset})} = I_{B1}R_f - I_{B2}R_3\left(1 + \frac{R_f}{R_1}\right)$$

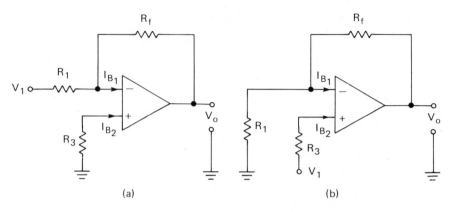

Figure 8.5 (a) Inverting configuration. (b) Non-inverting configuration.

With $1/\beta = 1 + (R_f/R_1)$,

$$V_{o(\text{offset})} = I_{B1}R_f - \frac{I_{B2}R_3}{\beta}$$

The output offset voltage due to the bias currents is minimized if $R_3 = R_1R_f/(R_1 + R_f)$. The preceding equation reduces to

$$V_{o(\text{offset})} = (I_{B1} - I_{B2})R_f$$

With $I_{OS} = I_{B1} - I_{B2}$,

$$V_{o(\text{offset})} = I_{OS}R_f$$

For the case where $R_3 = 0$, and since $I_B \approx I_{B1} \approx I_{B2}$

$$V_{o(\text{offset})} = I_BR_f$$

where

$$I_B = \frac{I_{B1} + I_{B2}}{2}$$

A typical op amp operating as an inverting or noninverting amplifier will have an output offset voltage due to the input offset voltage (V_{os}), input bias current (I_B), and input offset current (I_{os}) operating simultaneously. With all three conditions operating, the output offset is given by

$$V_{o(\text{offset})} = \frac{V_{os}}{\beta} + I_{B1}R_f - \frac{I_{B2}R_3}{\beta}$$

For the special case where $R_3 = R_1R_f/(R_1 + R_f)$,

$$V_{o(\text{offset})} = \frac{V_{os}}{\beta} + I_{os}R_f$$

For $R_3 = 0$,

$$V_{o(\text{offset})} = \frac{V_{os}}{\beta} + I_BR_f$$

To clarify the above theory and to give an idea of the range of values discussed, the following example is given.

Example 8.1

A National Semiconductor IC Op Amp, type LM101, has the following input parameters: $V_{OS} = 6$ mV, $I_{OS} = 200$ nA, and $I_B = 1.5$ μA. The amplifier has a gain of 10, with $R_f = 100$ kΩ, $R_1 = 10$ kΩ, and $R_3 = 9$ kΩ.

1. Determine the output offset voltage if the amplifier is to be used in its inverting configuration.
2. What is the output offset voltage if R_3 equals 0 Ω?

    Solution:

1.
$$\frac{R_f R_1}{R_f + R_1} = \frac{(100 \text{ k}\Omega)(10 \text{ k}\Omega)}{100 \text{ k}\Omega + 10 \text{ k}\Omega} = \frac{10^9}{1.1 \times 10^5} = 9 \text{ k}\Omega$$

With $R_f R_1/(R_f + R_1) = R_3 = 9$ kΩ

$$V_{o(\text{offset})} = \frac{V_{os}}{\beta} + I_{os}R_f; \qquad \beta = \frac{R_1}{R_1 + R_f} = \frac{10^4}{10^5 + 10^4} = \frac{1}{11}$$

$$= 6 \times 10^{-3}(11) + 200 \times 10^{-9}(10^5)$$

$$= 66.0 \text{ mV} + 20 \text{ mV} = \mathbf{86.0 \text{ mV}}$$

2. With $R_3 = 0$ Ω

$$V_{o(\text{offset})} = \frac{V_{os}}{\beta} + I_B R_f$$

$$= \frac{6 \times 10^{-3}}{1}(11) + 1.5 \times 10^{-6}(10^5)$$

$$= 66.0 \text{ mV} + 150 \text{ mV} = \mathbf{216.0 \text{ mV}}$$

Note that for a minimum offset output voltage, R_3 should be equal in value to R_f in parallel with R_1, or $R_3 = R_f R_1/(R_f + R_1)$.

8.3.2   Measurement of Input Offset Voltage (V_{OS}), Offset Current (I_{OS}), and Bias Current (I_B)

The circuit of Fig. 8.6 can be utilized to measure the three input parameters V_{OS}, I_B, and I_{OS}.

With switches S_1 and S_2 closed, measure the output offset voltage (V_{o1}). This voltage is V_{OS}. With switch S_2 closed and S_1 opened, determine the new output voltage V_{o2}. I_{B1} is determined by

$$I_{B1} = \frac{V_{o2} - V_{o1}}{10^6}$$

With switch S_1 closed and S_2 opened, measure the new output voltage V_{o3}. I_{B2} is determined by

$$I_{B2} = \frac{V_{o3} - V_{o1}}{10^6}$$

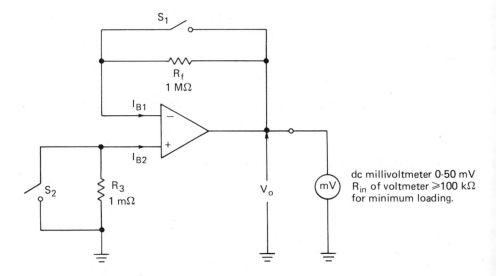

Figure 8.6 Circuit configuration for measuring dc offset voltage V_{OS}, input bias current I_B, input offset current I_{OS}.

The bias current I_B is found by

$$I_B = \frac{I_{B1} + I_{B2}}{2}$$

The input offset current I_{OS} is found by

$$I_{OS} = |I_{B1} - I_{B2}|$$
$$= \frac{V_{o2} - V_{o1}}{10^6} - \frac{V_{o3} - V_{o1}}{10^6} = \left| \frac{V_{o2} - V_{o3}}{10^6} \right|$$

8.3.3 Drift of Offset Voltage (αV_{OS}) and Current (αI_{OS}) as a Function of Temperature

An output dc voltage for zero input caused by input offset voltage (V_{OS}) and offset current (I_{OS}) in itself is not a serious problem since the output may be compensated for by injecting an equal and opposite signal at the summing junction. However, any tendency for the output voltage to drift due to temperature, time, or supply voltage variations would necessitate constant readjustment. The principal causes of drift are changes in offset voltage and current as a function of temperature.

For the inverting input amplifier, drift in V_{OS} and I_{OS} causes an output voltage drift, given by

$$\Delta e_o = \left| \frac{\alpha V_{OS}\, \Delta t}{\beta} \right| + |\alpha I_{OS}\, \Delta t\, R_f|$$

where Δe_o = output voltage drift,

αV_{OS} = average change in input offset voltage as a function of temperature in $\mu V/°C$,

αI_{os} = average change in input offset current as a function of temperature in $nA/°C$,

$\beta = R_1/(R_1 + R_f)$,

ΔT = change in temperature in °C.

An example will clarify how to determine drift in output voltage as a function of temperature.

Example 8.2

1. For an inverting dc amplifier with a gain of -10 ($R_f = 10\ k\Omega$, $R_1 = 1\ k\Omega$), with a αV_{OS} of 5 $\mu V/°C$ and αI_{os} of 10 $nA/°C$, determine the drift in output voltage for a temperature rise of 50°C.
2. Find the output voltage drift if the gain is -100 ($R_f = 10\ k\Omega$, $R_1 = 100\ \Omega$).

Solution:

1.
$$\beta = \frac{R_1}{R_1 + R_f} = \frac{1\ k\Omega}{10\ k\Omega + 1\ k\Omega} = \frac{1}{11}$$

$$\Delta e_o = \frac{\alpha V_{OS}\ \Delta t}{\beta} + \alpha I_{os}\ \Delta t\ R_f$$

$$= \frac{(5 \times 10^{-6})(50)(11)}{1} + (10 \times 10^{-9})(50)(10^4) = \textbf{7.75 mV}$$

2.
$$\beta = \frac{R_1}{R_1 + R_f} = \frac{100\ \Omega}{100\ \Omega + 10\ k\Omega} = 9.9 \times 10^{-3}$$

$$\Delta e_o = \frac{(5 \times 10^{-6})(50)}{9.9 \times 10^{-3}} + (10 \times 10^{-9})(50)(10^4)$$

$$= 25.3\ mV + 5\ mV = \textbf{30.3 mV}$$

Note: As the gain increases, the drift due to αV_{OS} correspondingly increases.

8.3.4   *Measurement of αV_{os} and αI_{os}*

The offset temperature drift may be measured utilizing the circuit of Fig. 8.6. With S_1 and S_2 closed, note the temperature and V_{o1}. Raise temperature a fixed amount (ΔT), and note the new output voltage V_{o2}. αV_{os} is found by

$$\alpha V_{os} = \frac{V_{o2} - V_{o1}}{\Delta t}$$

Open switch S_1 (S_2 closed) and determine output V_{o3}. Close S_1, open S_2, and measure output V_{o4}. Increase temperature by a known amount (ΔT) and measure a new V_{o3} and V_{o4}. αI_{os} is determined by

$$\alpha I_{os} = \frac{(V_{o3} - V_{o4})_{\text{new}} - (V_{o3} - V_{o4})_{\text{old}}}{10^6\ \Delta t}$$

8.3.5   Open Loop Voltage Gain (A_{VOL})

Open loop voltage gain A_{VOL} is defined as the ratio of the change in output voltage (v_o) for an input differential voltage ($v_i = v_+ - v_-$). Refer to Fig. 8.1.

As the open loop gain A_{VOL} of an amplifier decreases, its loop gain ($A_{VOL}\beta$) decreases, resulting in a deterioration in the drift, stability, input impedance, output impedance, and bandwidth. Refer to Section 8.2.2.

8.3.6   Measurement of the Open Loop Voltage Gain (A_{VOL})

Manufacturers of operational amplifiers usually specify the open loop voltage gain A_{VOL} for a dc input. A_{VOL} will be measured with an input ac signal operating at 5 Hz. This will simplify the measurement, and at this frequency the A_{VOL} measured will be approximately equal to A_{VOL} dc.

Prior to performing this measurement it is essential that the effect of the input offset parameters (V_{OS}, I_{OS}, I_B) be cancelled; otherwise the high amplification will cause the output voltage to saturate. With reference to Fig. 8.7,

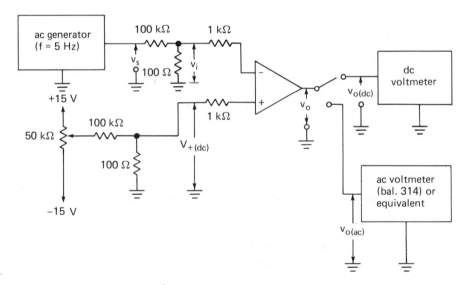

Figure 8.7   Measurement circuit for determining A_{vol} and 3-dB open loop bandwidth.

with a zero ac input voltage, the output is initially connected to a dc voltmeter, and the 50-kΩ potentiometer is adjusted for an output dc voltage of 0 ± 0.1 V dc.

The op amp output is connected to an ac voltmeter and the input ac signal is set to a sufficiently low level for an output ac signal less than one third of the op amp maximum output swing. Measure the output ac voltage (v_o) and input signal (v_s). With $A_{\text{VOL}} = v_o/v_i$, and since

$$v_i = \frac{v_s(100)}{100,100} \approx \frac{v_s(100)}{100,000} = \frac{v_s}{10^3}$$

then

$$A_{\text{VOL}} = \frac{1,000 v_o}{v_s}$$

The circuit of Fig. 8.7 may also be utilized to determine the open loop voltage gain at a particular frequency. This is accomplished by simply adjusting the input signal to the desired frequency and repeating the previous measurements.

8.3.7   Open Loop (3-dB) Bandwidth (BW_{OL}) and Unity Gain Bandwidth (f_t)

Another important parameter op amp manufacturers specify is the 3-dB open loop bandwidth BW_{OL} of the op amp. This is defined as the frequency where the open loop gain A_{VOL} drops 3 dB from its dc or low-frequency voltage gain. A drop of 3 dB in voltage gain is equivalent to the voltage gain decreasing to 0.707 of its original value. Refer to Fig. 8.8 for a plot of A_{VOL} vs. frequency for a typical op amp.

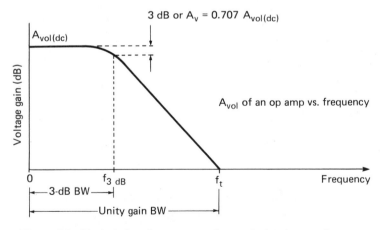

Figure 8.8   Typical plot of open loop voltage gain (A_{vol}) versus frequency.

Some manufacturers may also specify the unity gain bandwidth factor f_t, which is defined as the frequency at which the open loop voltage gain reduces to unity. Note that the unity gain bandwidth factor is exceedingly larger than the 3-dB bandwidth.

8.3.8   Measurement of BW_{OL} and f_t

The two bandwidth parameters BW_{OL} and f_t are measured with the circuit utilized for the open loop voltage gain with the exception of an oscilloscope to monitor the input and output voltages. Refer to Fig. 8.7. The test procedure is similar to measuring A_{VOL}, except one increases the input signal frequency until the output voltage drops to 0.707 of the output voltage at 5 Hz. The input signal amplitude should be held constant as the frequency is varied. The open loop 3-dB bandwidth (BW_{OL}) is the signal frequency for which the output falls to 0.707 of its output of 5 Hz. With the input voltage held constant, increase your signal frequency until the output voltage is equal to $v_s/1{,}000$. The unity gain bandwidth factor f_t is the signal frequency at which this occurs.

8.3.9   Slew Rate (S_r)

The previous bandwidth tests determined the frequency response of the operational amplifier under small-signal conditions. Many times the op amp must be utilized for large-signal conditions. A typical example is the op amp circuit of Fig. 8.9(a), which converts the input sinusoid into a square wave.

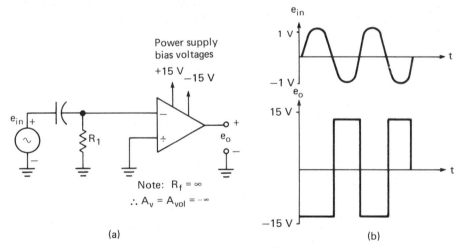

Figure 8.9   (a) Utilizing the op amp as a squarer, a circuit which converts a sinusoid into a square wave. (b) Squarer waveforms.

With no feedback, ($A_v = A_{VOL}$), the output will be driven until the amplifier saturates, which is approximately equal to the power supply dc bias voltage. Refer to Fig. 8.9(b). The maximum rate of change (volts per microsecond) of the output voltage when supplying rated output is defined as the slew rate.

Although both slew rate and bandwidth limit the maximum useful frequency of the amplifier, they produce different effects on a sinusoidal signal. A limited bandwidth will decrease the output voltage as a function of frequency with no distortion, while a finite slew rate affects the sinusoidal output by distorting its wave shape. The relationship between slew rate and maximum sinusoidal amplitude and frequency (with no distortion) is given by

$$S_r \geq 2\pi f e_p$$

where S_r = slew rate in $V/\mu s$,

f = sinusoidal frequency in MHz,

e_p = peak value of sinusoidal output in V.

An example will clarify the above theory.

Example 8.3

The output of an op amp is to be 2 V peak. If the slew rate is 5 $V/\mu s$, determine the maximum sinusoidal frequency for no distortion in the output signal.

$$S_r = 2\pi f e_p$$

or

$$f_{max} = \frac{S_r}{2\pi e_p} = \frac{5}{(2)(3.14)(2)} = 0.796 \text{ MHz or 796 kHz}$$

8.3.10   Measurement of Slew Rate (S_r)

The slew rate can be measured utilizing the inverting amplifier circuit of Fig. 8.10(a). The input is a low-frequency square wave (100 Hz) with an amplitude level such that the output will swing over its full range (saturate). Monitoring the output with an oscilloscope [shown in Fig. 8.10(b)], the slew rate is $\Delta V / \Delta t$.

8.3.11   Open Loop Input Impedance (R_{in})

The input impedance of an op amp consists of that impedance between the two input terminals and between each input and ground. Refer to Fig. 8.11. The impedance between the inverting and noninverting input is called the differential input impedance (R_{in}), with the impedance between both inputs and ground being referred to as the common mode impedance (R_c). With the op amp operating as an inverted amplifier, the input impedance will consist of R_{in} in parallel with $2R_c$. In a quality op amp, R_c is much larger than R_{in} and therefore is usually neglected.

The primary effect of R_{in} is to reduce the amplifier loop gain ($A_{VOL}\beta$) and consequently alter gain, accuracy and stability, BW, etc.

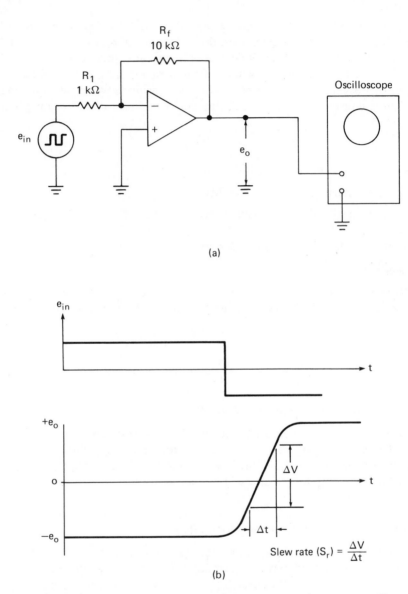

(a)

(b)

Figure 8.10   (a) Circuit for measuring the slew rate of an op amp. (b) Input and output voltages for above circuit in (a).

8.3.12   *Measurement of Differential Input Impedance (R_{in})*

The procedure which follows is practical only in measuring R_{in} for an op amp having a transistor input stage. Op amps with FET input stages have a R_{in}

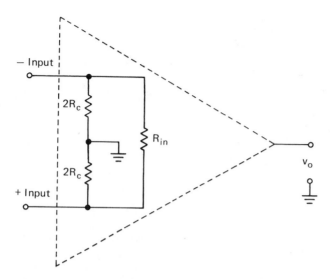

Figure 8.11 An op amp with its differential input impedance R_{in} the common mode impedance R_c.

on the order of thousands of megohms, and special test circuits are necessary to measure their input impedance.

The input impedance R_{in} may be measured with the circuit of Fig. 8.7, except that the two 1-kΩ resistors are omitted. The 50-kΩ potentiometer is initially adjusted for an output dc voltage of 0 ± 0.1 V. The signal generator voltage amplitude v_s is increased for an output V_{o1} of 1 V. Two equal resistors (R_x) are inserted where the 1-kΩ resistors were, and the new output voltage V_{o2} is noted. Utilizing the voltage divider rule and solving for R_{in}, we obtain

$$R_{in} = \frac{2R_x v_{o2}}{v_{o1} - v_{o2}}$$

Note: Two equal resistors were utilized instead of one such that any stray coupling from the output will generate equal signals at both inputs, cancelling each other.

8.3.13 *Open Loop Output Impedance (R_o)*

The open loop output impedance R_o forms a voltage divider with the load and feedback impedance and effectively attenuates the open loop voltage gain A_{VOL}. A reduction in A_{VOL} deteriorates the overall closed loop response of the amplifier.

8.3.14 *Measurement of Open Loop Output Impedance (R_o)*

The circuit of Fig. 8.7, slightly modified, may be used in measuring R_o. See Fig. 8.12. As in the measurement of A_{VOL} (Section 8.3.5), the 50-kΩ potentiom-

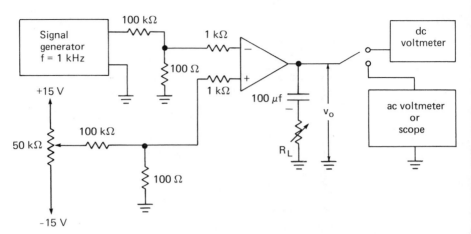

Figure 8.12 Circuit configuration for measuring the open loop output impedance R_o.

eter is initially adjusted for an output of 0 ± 0.1 V dc. The signal generator frequency is set to 1 kHz, with its input level adjusted for an ac output of 1 V.

A variable resistor R_L in series with a 100-μF capacitor is connected across the output and adjusted until the output drops to $\frac{1}{2}$ V. The resistance value of R_L, measured with an ohmmeter, is the output resistance of the op amp.

Note: It will be a student exercise to prove that $R_o = R_L$ value when v_o drops to one-half its original value.

8.3.15 Common Mode Rejection Ratio

An ideal op amp will amplify the difference signal between the inputs ($v_+ - v_-$) and will produce no output for a common mode voltage v_{cm}—that is, when both inputs are at the same potential. However, due to slightly different gains between the plus and minus inputs, an output will be produced for a common mode input voltage v_{cm}. This output due to the common mode input voltage v_{ocm} is generally referred to the input by dividing by the differential gain A_d and is called the common mode error voltage v_{ecm}:

$$v_{ecm} = \frac{v_{ocm}}{A_d}$$

where A_d is the differential voltage gain.

The common mode rejection ratio (CMRR) is then defined as the ratio of common mode voltage to common mode error voltage—that is,

$$\text{CMRR} = \frac{v_{cm}}{v_{ecm}} = \frac{v_{cm} A_d}{v_{ocm}}$$

Manufacturers of op amps usually express this term in decibels:

$$\text{CMRR}_{\text{dB}} = 20 \log \left(\frac{v_{cm}}{v_{ecm}}\right)$$

Errors due to a finite common mode rejection ratio arise only when the amplifier is used as a noninverting or differential amplifier. An example will clarify this.

Example 8.4

Given a differential amplifier with a differential gain A_d of 10,000 and a CMRR of 74 dB, determine the output voltage due to a 1.0-V common mode signal. Refer to Fig. 8.13.

$$\text{CMRR} = 74 \text{ dB} = 20 \log \left(\frac{v_{cm}}{v_{ecm}}\right)$$

$$\log \left[\frac{v_{cm}}{v_{ecm}}\right] = \frac{74}{20} = 3.7$$

$$\frac{v_{cm}}{v_{ecm}} = 5{,}012$$

$$v_{ecm} = \frac{v_{cm}}{5{,}012} = \frac{1.0}{5{,}012}$$

$$v_{ocm} = A_d v_{ecm} = \frac{(1)10^4}{5{,}012} = \textbf{2.0 V}$$

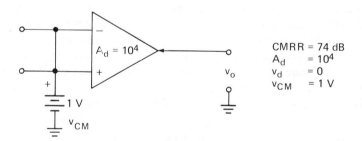

CMRR = 74 dB
A_d = 10^4
v_d = 0
v_{CM} = 1 V

Figure 8.13

8.3.16 Measurement of Common Mode Rejection

Refer to the circuit of Fig. 8.14 for measuring the common mode rejection ratio. As in the previous measurements, the 50-kΩ potentiometer is initially adjusted for an output of 0 ± 0.1 V dc. The signal generator frequency is set to 1 kHz, with an amplitude of 0.5 V. Measuring the ac output voltage, v_o, one determines the CMRR by

$$\text{CMRR} = \frac{R_f}{R_1} \frac{v_s}{v_o} = \frac{100 v_s}{v_o} = \frac{(100)(0.5)}{v_o} = \frac{50}{v_o}$$

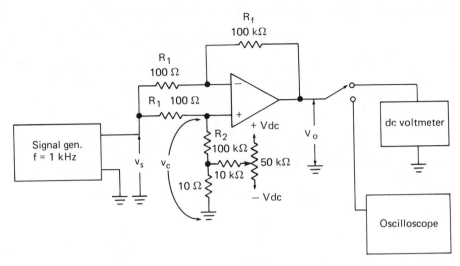

Figure 8.14   Circuit for measuring the CMRR.

8.4   DIGITAL INTEGRATED CIRCUITS

The most common families of digital logic elements available today are the transistor-transistor logic (TTL), the emitter-coupled logic (ECL), and the complementary metal-oxide semiconductor logic (CMOS). In this chapter we shall primarily discuss the characteristics and measurements of the TTL logic element, because it is presently the most widely used and has the largest selection of logic configurations that operate at reasonably high speeds and have low power requirements. A thorough knowledge of TTL characteristics and measurements will enable the student to interpret and apply the manufacturers' specifications for other types of digital devices.

8.5   TRANSISTOR-TRANSISTOR LOGIC (TTL)
       INTEGRATED CIRCUIT

In 1964, Texas Instruments introduced a TTL logic line which they designated semiconductor network series 54 (SN54), intended primarily for military applications. These devices were designed to operate under high temperature variations with extreme reliability. Texas Instruments later introduced a low-cost industrial version designated SN74. The series SN54/74 of TTL devices eventually evolved into four divisions: standard (SN54/74), high speed (SN54H/74H), low power (SN54L/74L), and Schottky diode clamped (SN54S/74S). All four families are compatible and will interface directly with one another. Refer to Table 8.2 for typical characteristics of the four TTL families and ECL and CMOS devices.

Table 8.2

Comparison of Integrated Circuit Logic Families with Respect to
Propagation Time, Power Dissipation, and Fan-out

IC Family	Propagation Time (ns)	Power Dissipation Per Gate (mW)	Max. Fan-out
TTL 54/74	10	10	10
TTL 54L/74L	33	1	10
TTL 54H/74H	5	23	10
TTL 54S/74S (Schottky)	3	20	10
ECL	2	50	25
CMOS	50-100	10^{-5}	Very high

In the section which follows we shall discuss characteristics and measurements of an SN54/74 TTL NAND gate. It is important to realize that the characteristics and measurement techniques described for the NAND gate are applicable to other TTL configurations such as the inverter; AND, NOR, and AND-OR gates, etc.

The electrical characteristics of a 54/74 TTL NAND gate are given in Table 8.3. Each term listed will be defined with a circuit shown for measuring the specified values. All measurements made will be worst-case tests and will be performed at room temperature.

8.5.1   Theory of Operation of the TTL (SN54/74) NAND Gate

A four-input NAND gate logically shown in Fig. 8.15(a), may be implemented utilizing a TTL SN54/74 IC as shown in Fig. 8.15(b). The circuit operates as follows. With all four inputs at a high input voltage (logical 1 level) transistor Q_1 is reversed biased, with its base current diverted from the emitter of Q_1 to the collector of Q_1, causing Q_2 to conduct. This current through Q_2 biases Q_3 on, which effectively reduces the emitter impedance of Q_2, which further increases the current flow through Q_2. The collector voltage of Q_2 falls, turning off Q_4 and saturating Q_3. The output voltage is approximately 0 V or in a logical 0-level state.

With one or more of the emitter inputs of Q_1 at a low input voltage (logical 0 level), transistor Q_1 will conduct. Since the collector of Q_1 is connected to the base of Q_2, and with a large reverse base current not possible, Q_1 saturates. With Q_1 saturated and supplying a low-impedance path for any

Table 8.3

Electrical Characteristics of a TTL NAND Gate

PARAMETER		TEST CONDITIONS	MIN	TYP	MAX	UNIT
V_{IH}	Logical 1 input voltage required to ensure a logical 0-level output	V_{CC} = Min	2			Volts
V_{IL}	Logical 0 input voltage required at input terminal to ensure logical 1-level at output	V_{CC} = Min			0.8	Volts
V_{OH}	Logical 1 output voltage	V_{CC} = Min, V_{In} = 0.8 V I_{Load} = 400 μA	2.4	3.3		Volts
V_{OL}	Logical 0 output voltage	V_{CC} = Min, V_{In} = 2 V I_{Sink} = 16 ma		0.2	0.4	Volts
I_{IH}	Logical 1-level current (each input)	V_{CC} = Max V_{In} 2.4 V			40	μA
I_{IL}	Logical 0-level input current (each input)	V_{CC} = Max, V_{In} = 0.4 V			1.6	mA
I_{OS}	Short circuit output current	V_{CC} = Max SN5400	20		55	mA
		SN7400	18		55	mA
I_{CCL}	Logical 0-level supply current	V_{CC} = Max, V_{In} = 5 V		12	22	mA
I_{CCH}	Logical 1-level supply current	V_{CC} = Max, V_{In} = 0 V		4	8	mA

Switching characteristics, V_{CC} = 5 V, Fan - out = 10						
PARAMETER		TEST CONDITIONS	MIN	TYP	MAX	UNIT
t_{PHL}	Propogation delay time for output to go from logical 1-level to 0-level	C_L 15 pF, R_L = 400 Ω		7	15	ns
t_{PLH}	Propogation delay time for output to go from logical 0-level to 1-level	C_L = 15 pF, R_L = 400 Ω		11	22	ns

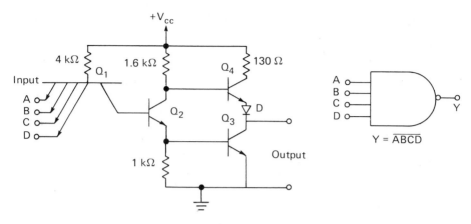

Figure 8.15 (a) Logical symbol of a NAND gate. (b) Electrical schematic diagram of a SN 54/74 TTL four-input NAND gate.

stored charge at the base of Q_2, transistor Q_2 will quickly turn off. With Q_2 cut off, its collector is at a potential equal to V_{cc}, with its emitter at zero voltage. Transistor Q_4 will conduct while Q_3 is cut off, resulting in a high output voltage or a logical 1 level.

Transistors Q_3 and Q_4 make up what is referred to as a totem-pole or active pull-up output. Their purpose is to provide a low driving source impedance for capacitive circuits, resulting in a fast switching time for the output going from logical 0 to 1 or 1 to 0 levels. With the output a logical 1 state, its output impedance is low since Q_4 appears as an emitter follower source, driving current into the load. For a logical 0 state, Q_4 is cut off with Q_3 saturated, resulting in a low saturation resistance for any capacitive loads.

8.5.2 TTL SN54/74 NAND Gate Source and Sink Currents

For a 54/74 NAND gate with all inputs a logical 1-level voltage (V_{IH}), the base to emitter junction will be reverse biased, cutting off transistor Q_1. Under these conditions we would assume that the input current is zero, but in actuality a small leakage current of 40 μA (max.) may flow. Refer to Fig. 8.16. With the 54/74 NAND gate having a fan-out of 10, a gate must be capable of driving 10 other gates, supplying a source current of (10 times 40 μA) 400 μA.

With the gate having a logical 0-level input voltage, transistor Q_1 is saturated. Under these conditions, the maximum current flow from collector to emitter (Q_1) will be 1.6 mA. Since a gate with a logic 0-level output is capable of driving 10 identical gates (fan-out = 10), its output transistor Q_3, which is also saturated, should be capable of sinking a current of (10 times 1.6 mA) 16 mA. Refer to Fig. 8.17.

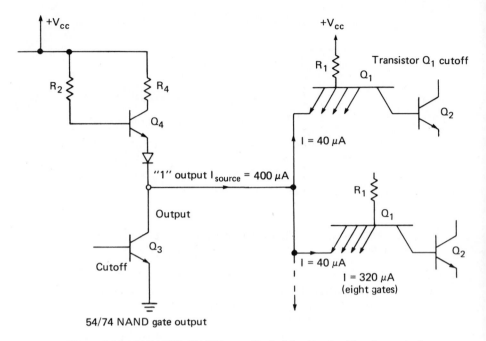

Figure 8.16 54/74 TTL NAND gate (logical 1 output) with a fan-out of 10.

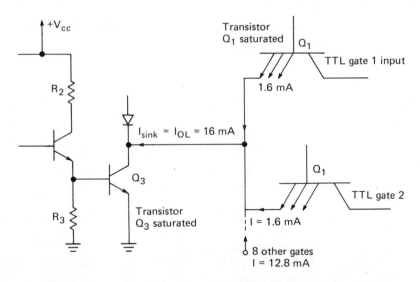

Figure 8.17 54/74 TTL NAND gate (logical 0 output) driving 10 other gates.

8.6 MEASUREMENT OF TTL SN54/74 NAND GATE CHARACTERISTICS

8.6.1 Measurement of V_{IH} and V_{OL}

V_{IH} is defined as the minimum 1-level input voltage required to guarantee a 0-level output. The maximum 0-level output is defined as V_{OL}. With reference to Fig. 8.15(b), with the output to be a logical 0, transistor Q_3 must be saturated. A worst-case combination of input, output, and V_{cc} must be determined and must still ensure a 0-level output. With the transistor (Q_3) collector current $I_c = \beta I_B$, the transistor will go out of saturation either by decreasing its base current or increasing its collector current. With I_B being a function of V_{cc}, a worst-case test results if V_{cc} is set to its minimum allowable voltage. For the 54 series $V_{cc(min)}$ is 4.50 V dc, and for the 74 series $V_{cc(min)}$ is 4.75 V dc.

It was previously mentioned that a gate with a logic 0 output must be capable of sinking 10 other gates, or of sinking a maximum current of 16 mA. This will be another worst-case condition for increasing I_c such that the transistor goes out of saturation. A worst-case test will also result if all inputs are held at the minimum logical 1 voltage, which is 2.0 V.

V_{IH} is therefore tested as follows: (1) V_{cc} is set to a minimum, (2) the output sink current is to be 16 mA, and (3) all inputs are connected together with a minimum input voltage of 2.0 V. The maximum output voltage for the 0 logic output is to be 0.4 V.

A test circuit for determining V_{IH} and V_{OL} is shown in Fig. 8.18. R_{in} adjusts the input voltage, while R_o is varied for the proper sink current of 16 mA.

V_{IH} measurement. Initially R_{in} is varied such that V_{in} is a maximum (logical 1 level), with R_o then adjusted until I_{sink} is 16 mA. The input voltage

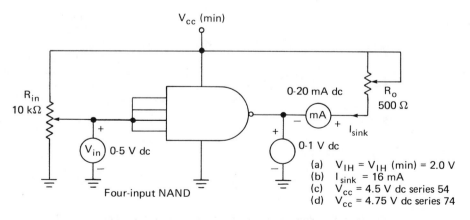

Figure 8.18 Worst-case test circuit for measuring V_{IH} and V_{OL}.

219

V_{in} is then decreased until V_o is 0.4 V, with R_o correspondingly adjusted for a sink current of 16 mA. The minimum input voltage which will just result in 0.4 V out at a 16-mA sink current is V_{IH}. This voltage, according to the manufacturer, should be a minimum of 2.0 V.

V_{OL} **measurement.**   With reference to Fig. 8.18, R_{in} is adjusted for an input of 2.0 V, while R_o is varied for a 16-mA sink current. The measured output voltage (V_o) is V_{OL} and according to the manufacturer's specification should be a maximum of 0.4 V.

8.6.2   Measurement of V_{IL} and V_{OH}

V_{IL} is the maximum 0-level input voltage required to guarantee a 1-level output. The minimum 1-level output is defined as V_{OH}.

For the output to be at a logical 1 level, Q_3 is cut off, with Q_4 conducting and supplying a current of 40 μA/gate [see Fig. 8.15(b)]. With the device having a fan-out of 10, a worst-case test of V_{IL} and V_{OH} is made with the gate supplying a 400-μA output current. Since a small V_{cc} will also cause a small v_o, these parameters are determined with V_{cc} set to a minimum. With a logical 0 level applied to the input, and where that voltage is required to hold the output transistor in its off state, a worst-case test results, with all other inputs returned to V_{cc}.

Measurement of V_{IL}.   With reference to Fig. 8.19, R_{in} is initially set for V_{in} equal to 0 V, resulting in the output a logical 1 level. R_o is then adjusted for a load current of 400 μA. V_{in} is then increased until V_o is reduced to a minimum of 2.4 V, with a load current of 400 μA. The maximum input voltage for a V_o of 2.4 V at 400-μA load current is V_{IL}. According to the manufacturer's specification, V_{IL} should be a maximum of 0.8 V.

Measurement of V_{OH}. V_{OH} is measured by adjusting the R_{in} of Fig. 8.19 for an input of 0.8 V and varying R_o for a load current of 400 μA. The

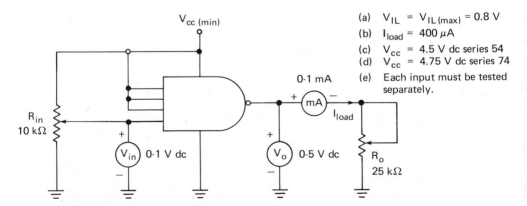

Figure 8.19   Worst-case test circuit for measuring V_{IL} and V_{OH}.

measured output voltage V_o is V_{OH} and according to the manufacturer's specification should be a minimum of 2.4 V.

8.6.3 Measurement of I_{IL}

The maximum input current that flows when the input is at a logical 0 level (0.4 V max.) is defined as I_{IL}. With I_{in} increasing as V_{cc} is increased, a worst-case test of this parameter is performed with V_{cc} set to its maximum value. With the exception of the input under test, all unused inputs are connected to V_{cc} to maximize any contribution of these inputs to I_{IL}.

This test is performed to screen out those devices with an I_{IL} of sufficient magnitude to cause a V_{OL} greater than 0.4 V when driving 10 such inputs. Since a V_{OL} logical 0 level of 0.4 V max. is tested while sinking a current of 16 mA, I_{IL} for a single device is to be a maximum of 1.6 mA when V_{in} is 0.4 V. Note that the current I_{IL} flows out of the gate, not into it.

With reference to Fig. 8.20, R_{in} is adjusted for an input voltage of 0.4 V. The milliammeter reading is I_{IL}.

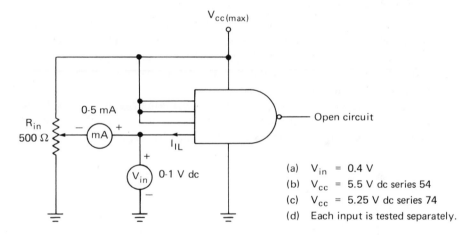

Figure 8.20 Worst-case test circuit for measuring I_{IL}.

8.6.4 Measurement of I_{IH}

The input current that flows when the input is at a logical 1 level (I_{IH}) is defined as I_{IH}. With the input a logical 1 level [Fig. 8.15(b)], Q_1 is back-biased, and a small leakage current, which is referred to as I_{IH}, flows.

A worst-case test results with V_{cc} set to its maximum and all unused inputs grounded. An acceptable device should have a maximum input current of 40 μA, with a V_{in} (equal to V_{IH}) of 2.4 V.

I_{IH} is measured utilizing the circuit of Fig. 8.21. R_{in} is adjusted for an input voltage of 2.4 V, with the resulting current flow being I_{IH}.

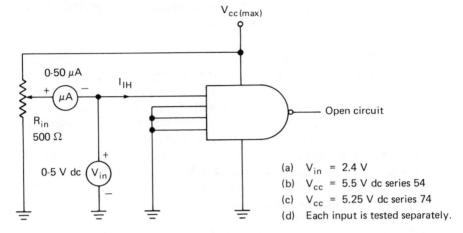

Figure 8.21   Worst-case test circuit for measuring I_{IH}.

8.6.5   Measurement of I_{os}

The maximum output current which will flow when its output is a logical 1 level and shorted to ground is referred to as I_{os}. This test checks the value of the current limiting resistor R_4 [Fig. 8.15(b)] and the proper operation of transistor Q_4 and diode D. A worst-case test results when V_{cc} is set to maximum value and all inputs are at zero voltage.

Refer to the test circuit of Fig. 8.22 for determining I_{os}. According to the manufacturer's specification, a satisfactory I_{os} is a current flow between limits of 20–55 mA for the 54 series and 18–55 mA for the 74 series of devices.

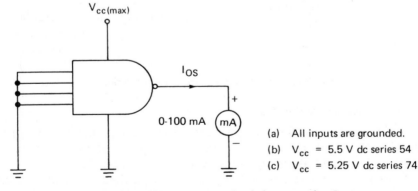

Figure 8.22   Worst-case test circuit for measuring I_{os}.

8.6.6   Measurement of I_{CCL} and I_{CCH}

I_{CCL} is defined as the power supply current flowing when all individual outputs of the integrated circuit are at a logical 0 level. The power supply current when the output is at a logical 1 level is defined as I_{CCH}. These tests are performed with V_{cc} set to a maximum, and they permit us to estimate the total power supply capacity required.

With reference to Fig. 8.23, R_{in} is adjusted for an input of 5.0 V, with the resulting current flow being a measure of I_{CCL}. With V_{in} set to 0 V, the measured current is I_{CCH}. The above measurements determine the current flow per gate. For an IC containing N gates, the total power supply current is $N \times I_{CCL}$ or $N \times I_{CCH}$.

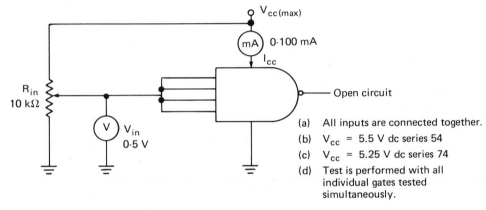

Figure 8.23   Worst-case test circuit for measuring I_{CCL} and I_{CCH}.

8.6.7   *Switching Speed Measurements t_{PHL}, t_{PLH}, and t_{PD}*

The switching time parameters to be determined for the 54/74 TTL devices are (1) the propagation delay time t_{PHL}, defined as the time required for an IC gate output to go from a logical 1 level to a logical 0 level; (2) t_{PLH}, the propagation time for the output to go from a logical 0 level to a logical 1 level; and (3) t_{PD}, which is simply the numerical average of t_{PHL} and t_{PLH}. These measurements are important in that they determine the speed of the gate operation and place an upper limit on the clock frequency that can be used.

See Fig. 8.24 for the circuit for testing these switching times. The diodes, resistor R_L, and capacitor C_L are used to approximate the loading of 10 TTL gate inputs. The pulse generator should be set for a pulse amplitude of 3.5 V, a pulse width of 0.5 μs, and a pulse repetition rate of 1 MHz. With the propagation delay being as low as 7 ns, the dual trace oscilloscope should have a vertical amplifier with a frequency response greater than 100 MHz. See Fig. 8.25 for typical gate waveforms. With a C_L of 15 pF, an SN54/74 gate should have a maximum of 15 ns with a t_{PLH} of 22 ns. The average propagation delay time, t_{PD}, is the numerical average of t_{PHL} and t_{PLH}.

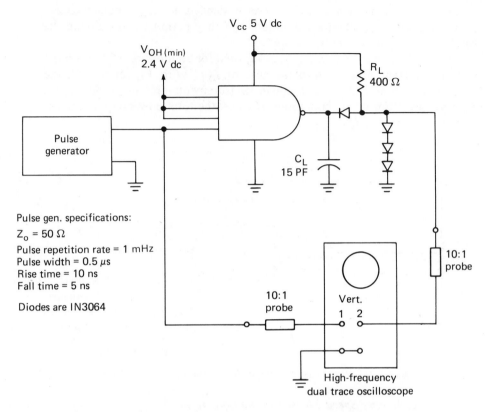

Figure 8.24   Test set up for measuring t_{PHL} and t_{PLH}.

8.7 SUMMARY

Integrated circuits can be divided into two general classes, linear and digital. The operational amplifier, being the most useful and popular linear IC, becomes a powerful and useful electronic device with negative feedback. While negative feedback reduces the open loop gain, all other parameters such as input and output impedance, bandwidth, offset voltage, and drift are altered such that an op amp approaches an ideal amplifier. Definitions and measurements of the following operational amplifier characteristics have been presented: open loop gain, bandwidth, input and output impedance, offset (voltage, current, and drift), slew rate, and common mode rejection ratio.

To help the student understand the operation of a digital IC, we have presented characteristics, definitions, and measurements of the TTL 54/74 IC gate: V_{IH}, V_{OL}, V_{IL}, V_{OH}, I_{IL}, I_{IH}, I_{OS}, I_{CCH}, I_{CCL}, t_{PHL}, and t_{PLH}.

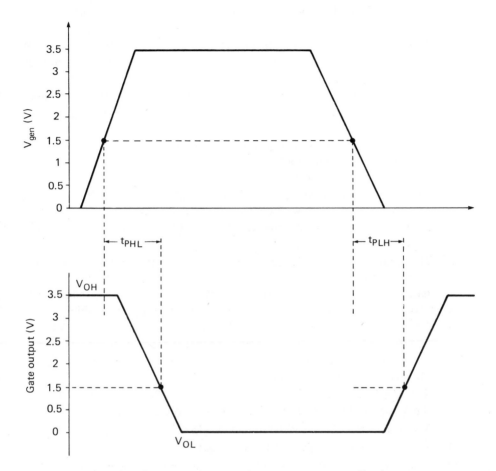

Figure 8.25 Voltage wave forms of TTL gate output vs. V_{gen} input.

REVIEW QUESTIONS

1. The op amp circuit of Fig. 8.5(a) ($R_1 = 100 \ \Omega$, $R_f = 10 \ \text{k}\Omega$, $R_3 = 100 \ \Omega$) has an open loop gain A_{VOL} of 10,000.
   a. Calculate the voltage gain A_v assuming the op amp is ideal.
   b. Determine the actual gain.
2. The loop gain ($A_{\text{VOL}}\beta$) of the op amp circuit shown in Fig. 8.5(a) equals 100. What is the open loop voltage gain A_{VOL} of the amplifier? $R_1 = 100 \ \Omega$, $R_f = 10 \ \text{k}\Omega$, and $R_3 = 100 \ \Omega$.
3. An amplifier was found to have a bandwidth of 100 kHz with an output impedance of 1 kΩ. Negative feedback is introduced with a loop gain

$(A_{VOL}\beta)$ of 50. Determine the bandwidth and output impedance with feedback.

4. The op amp circuit of Fig. 8.5(a) ($R_1 = 1$ kΩ, $R_f = 100$ kΩ, $R_3 = 909$ Ω) has an input offset voltage V_{OS} of 20 mV. Determine the output offset voltage.

5. An IC op amp has the following input characteristics: $V_{OS} = 10$ mV, $I_{OS} = 5$ μA, and $I_B = 20$ μA. The op amp is utilized as an inverting amplifier with $R_f = 50$ kΩ, $R_1 = 1$ kΩ, and $R_3 = 980$ Ω.
    a.   Determine the voltage gain of the amplifier.
    b.   Find the output offset voltage.
    c.   If $R_3 = 0$ Ω, find the output offset voltage.

6. The op amp circuit of Fig. 8.5(a) ($R_1 = 100$ Ω, $R_f = 10$ kΩ, $R_3 = 100$ Ω) has an input offset voltage drift of 10 μV/$^\circ$C, with an offset current drift of 50 nA/$^\circ$C. Determine the drift in output voltage for a temperature rise of 50°C.

7. The open loop gain (A_{VOL}) of an op amp amplifier is 100,000. The open loop gain dropped to 70,700 at a frequency of 100 Hz and fell to unity gain at a frequency of 1 MHz. Determine the open loop bandwidth (BW$_{OL}$) and the unity gain bandwidth factor (f_t) of the amplifier.

8. The slew rate S_r of the op amp in Fig. 8.5(a) was measured as 20 V/μs. Determine the maximum peak input voltage for a distortion-free output sinusoid. The frequency of the sinusiod is 1 kHz. $R_f = 10$ kΩ, $R_1 = 100$ Ω, and $R_3 = 100$ Ω.

9. The common mode rejection ratio of an amplifier is 100 dB. If V_{CM} is 5 mV, determine the output voltage for a differential gain of 5,000.

10. What does the "S" stand for in a 54S/74S TTL gate?

11. Define V_{IH} and V_{OL} for a TTL 54/74 gate.

12. What are I_{CCL} and I_{CCH} of a TTL gate? Why are the measurements made?

13. What determines the switching speed of a TTL 54/74 gate?

14. The propagation delay t_{PHL} is measured as 12 ns, with t_{PLH} measured as 14 ns. Determine the average propagation delay t_{PD}.

9

HIGH-FIDELITY AUDIO SYSTEMS AND TESTING

9.1 INTRODUCTION

Investigation of audio-frequency systems involves the study of sound from 20 to 20,000 Hz. Audio-frequency systems are used in broadcasting (the reproducing system) and in high fidelity. The main emphasis in the sections that follow will be on pickups or phonograph cartridges, amplifiers, and loudspeakers. Distortion and frequency response will be the topics of concern. We shall determine how distortion and frequency response can be measured. The newest innovation, the spectrum analyzer, will be used in amplifier studies. This will be followed by an explanation of terms (SPL, dB, etc.) used in sound studies. We shall then proceed to explain how the vu meter works and shall also discuss other audio instruments. We shall conclude the chapter by discussing hi-fi component ratings.

9.2 HIGH-FIDELITY SYSTEMS—CONCEPTUAL REMARKS

The ideal reproducing system should have the ability to reproduce music or speech faithfully over the entire audio spectrum. A reproducing system consists of the pickup, amplifiers, and the speaker system. The pickup discussed may be a stereo, monophonic, or quadraphonic cartridge. For the cartridge to play back without objectionable distortion, three requirements must be met. First, sensitivity to distortion must be minimized. Second, the cartridge output must be a linear function of stylus velocity (or amplitude). Finally, the cartridge must have a sufficiently low mechanical impedance* at the stylus, symmetrical for all directions of drive, to avoid distortion due to mistracking at low vertical stylus forces.

There are two methods for measuring distortion in cartridges. In the harmonic method, a single sinusoidal signal is applied to the cartridge under test. The output is then observed for frequency components which are integral multiples of the input signal. In the intermodulation method, two sinusoidal signals are applied. The distortion is determined from the amplitudes of the output components which are not harmonically related to either of the input signals. These types of distortion are applicable for study of the amplifier and speaker system also.

High-fidelity reproduction, in general, depends on the equipment being free from faults. After a sound arrives at a microphone pickup, its quality can be impaired by the same distortions that affect speech intelligibility.

A block diagram of a simplified single-channel high-fidelity system is shown in Fig. 9.1 The transducer or pickup (phonograph cartridge) transforms mechanical energy into electrical energy and is designated as the pickup. The signal from the transducer is amplified via a preamplifier-power amplifier and fed into the loudspeaker system consisting of one or more *woofers* and *tweeters*, and a crossover filter. The loudspeaker is the electromechanical system which changes electrical energy to mechanical energy for the high-fidelity enthusiast to hear. The weakest links in a high-fidelity system are the input device, the transducer (pickup), and the output device, the loudspeaker system. The input signal for a high-fidelity system may also be from a tuner(AM-FM) or a tape deck.

9.3 DISTORTION CONCEPTS AND MEASUREMENTS

9.3.1 Phonograph Cartridge Analysis†

Cartridge distortion [1] and losses can be classified as follows:

  1. Tracing distortion, which exists in monophonic and stereo cartridges because there exists a difference in shape between the recording and reproducing stylus. Such distortion depends on recorded wave amplitude, recorded wavelength, and stylus tip radius.

  2. Pinch distortion, which is measurable and will be discussed in Section 9.3.1.1.

  3. Translation loss, which is the difference between the reproduced levels

---

*Mechanical impedance is defined as the ratio of unit force to velocity.

†Section 9.3.1 is based on P. Kantrowitz, "Reproduction Distortion—Its Measurement and Influence on Stereo Phonograph Cartridge Performance," *Journal of the Audio Engineering Society*, Vol. 9, No. 2 (April 1961), pp. 115–120.

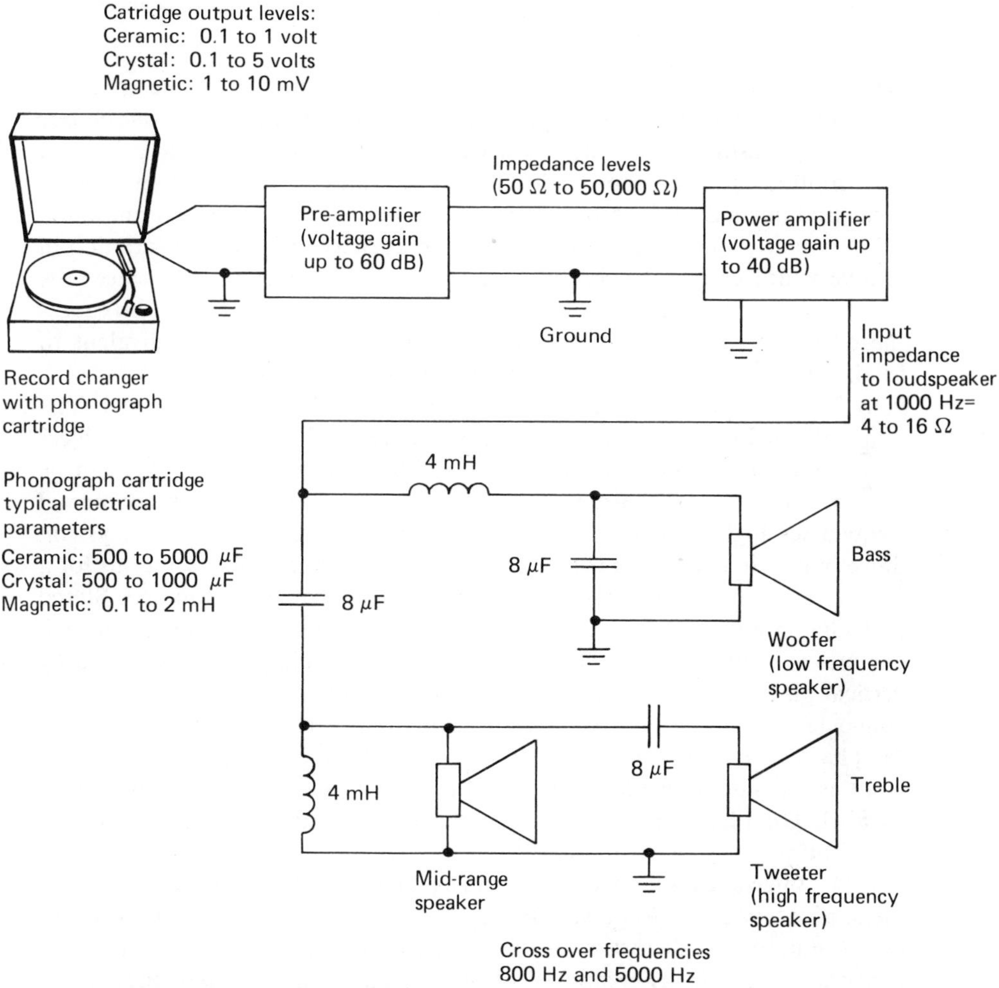

Catridge output levels:
Ceramic: 0.1 to 1 volt
Crystal: 0.1 to 5 volts
Magnetic: 1 to 10 mV

Impedance levels
(50 Ω to 50,000 Ω)

Pre-amplifier
(voltage gain
up to 60 dB)

Power amplifier
(voltage gain up
to 40 dB)

Ground

Input
impedance
to loudspeaker
at 1000 Hz=
4 to 16 Ω

Record changer
with phonograph
cartridge

Phonograph cartridge
typical electrical
parameters
Ceramic: 500 to 5000 μF
Crystal: 500 to 1000 μF
Magnetic: 0.1 to 2 mH

4 mH

8 μF

8 μF

8 μF

Bass

Woofer
(low frequency
speaker)

8 μF

4 mH

Treble

Mid-range
speaker

Tweeter
(high frequency
speaker)

Cross over frequencies
800 Hz and 5000 Hz

Figure 9.1   Single-channel high fidelity system. Note: voltage gain (in dB)
$= 20 \log_{10} \dfrac{\text{output voltage}}{\text{input voltage}}$. If the input voltage $= 1$ mV, the gain $= 60$ dB,
and the output voltage $= 1$ V.

at two different but equally modulated points of a record and can be measured by a specially designed test record or the CBS-STR-100 test record.

4. Scanning losses, which occur at high frequencies, usually above
10 kHz, causing a modulation wavelength effect. This loss involves the effect
of the finite size of the stylus-groove contact surface upon the force transmitted to the stylus. This phenomenon occurs when the modulation wavelength

becomes comparable to the surface dimensions. The variable component of the force between the stylus and the record will be reduced by partial cancellation of modulation effects.

9.3.1.1 Pinch distortion. An undesirable vertical motion of the spherical reproducing stylus tip is generated upon playback of lateral or monophonic recording because a relatively flat face stylus is used in the recording process. This motion is called the pinch effect. The width of the resultant groove, measured normal to the groove center line, varies and is widest at minimum velocity points and narrowest at maximum velocity points. This variation in groove width causes the reproducing stylus to rise and fall as it traces the groove. Thus, the presence of lateral motion plus vertical pinch motion causes the output of a stereo reproducer to contain a total signal equivalent to fundamentals plus second-order harmonic distortion terms.

9.3.1.2 Distortion and pinch effect. Listening tests indicate that total harmonic distortion is discernible at values between 7% and 10% and is objectionable for a frequency range of from 400 to 4,000 Hz. Little value can be obtained from total harmonic distortion measurements at fundamental frequencies higher than one-half the upper frequency limit of the system. Since our main concern is with second-order harmonic distortion terms for a 15-kHz system, fundamentals beyond 8kHz need not be considered. Stereo cartridges are designed so that the signals caused by vertical motion in each channel are out of phase. Therefore, second harmonic distortion caused by vertical pinch motion will be reduced considerably if the stereo channels are connected in parallel. A reduction in stylus radius results in a reduction of total harmonic distortion. In addition, the total harmonic distortion measurement is considerably reduced when the stereo channels are connected in parallel. Total harmonic distortion in many stereo cartridges can be reduced $2\frac{1}{2}$ to 3 times as a result of paralleling.

9.3.1.3 Intermodulation measurements as a measure of linearity. Measurement of intermodulation distortion by the SMPTE* method is necessary to evaluate nonlinear distortion at low frequencies. The SMPTE* method [3, 4] consists of test signals composed of one high and one low frequency which have been combined into a linear network.

The low-frequency signal is several decibels greater than the high-frequency signal. The combined signal is applied to the device under test, and the modulation of the high frequency by the low one in the output signal is observed. To indicate the degree of nonlinearity, the magnitude of the modulation envelope is compared to the magnitude of the original high-frequency portion of the test signal and expressed as a percentage. The RCA 12-5-39, 78-r/min, 12-in. monophonic intermodulation test record can be used as a source for checking stereo cartridges. The peak velocity of the 400-

*SMPTE refers to the Society of Motion Pictures and Television Engineers.

Hz signal level varies in eight approximately logarithmic steps from 4.3 to 27 cm/s, permitting investigation of these recording levels.

A comparison of intermodulation distortion percentages for different cartridges, all operating at the same vertical stylus force and at high levels of recorded velocity, indicates that intermodulation distortion generally decreases as cartridge compliance increases. There is a velocity level below which distortion for a cartridge is a minimum and does not change with vertical stylus force. Figure 9.2 shows typical intermodulation distortion as a function of stylus force measured by the SMPTE method for the stereo cartridge. Figure 9.3 compares the distortion for two different cartridges of this type with lateral compliances of 3.5 and 1.8×10^{-6} cm/dyn at a vertical stylus force of 4 g. A vertical stylus force of 4 g was chosen for this compari-

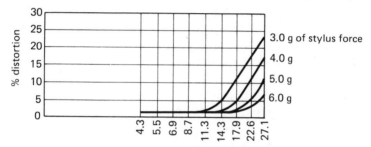

Figure 9.2 Intermodulation distortion (SMPTE) v. peak velocity at various vertical stylus forces for a stereo phonograph cartridge. Note: figures 9.2–9.5 have been reproduced by permission from "Reproduction Distortion—Its Measurement and Influence on Stereo Phonograph Cartridge Design," by Philip Kantrowitz, *Journal of Audio Engineering Society*, Vol. 9, No. 2, April 1961, pp. 115–120.

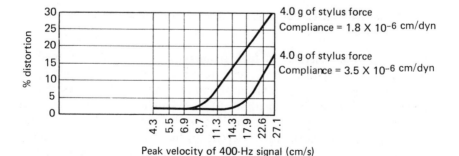

Figure 9.3 Intermodulation distortion (SMPTE) v. peak velocity for a stereo phonograph cartridge with a compliance of 3.5×10^{-6} cm/dyn and for an experimental unit with reduced compliance.

son because it is associated with an intermodulation distortion of less than 10% at 20 cm/s. This would indicate that a vertical stylus force of 4 g should be satisfactory for this cartridge, provided recording hertz levels at 400 Hz do not exceed 20 cm/s. As shown in Fig. 9.3, the cartridge with the lower compliance has considerably more distortion beyond 10 cm/s. As a measure of distortion in the record reproduction process, the SMPTE method is satisfactory for the evaluation of low-frequency tracking ability and linearity with large amplitude excursions.

9.3.1.4 Necessity for other distortion measurements.   Listening tests using a variety of stereo cartridges indicated that although the cartridges chosen had similar frequency response the sound of high-frequency musical passages varied. Thus, it would be desirable to measure high-frequency nonlinear distortion, a factor correlated with listening dissimilarities. Nonlinear distortion near the upper frequency limit of the system may be obtained by measurement of intermodulation distortion by the CCIF* or difference frequency method.

See Fig. 9.4 in regard to audible distortion as a function of frequency.

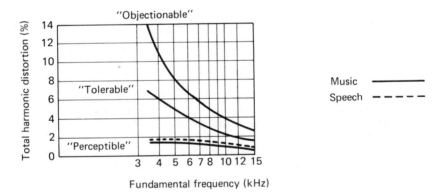

Figure 9.4   Audible distortion as a function of frequency (according to Olson [5]).

9.3.1.5 Distortion tolerable at high frequencies.   A search was made to determine how much distortion at high frequencies could be tolerated by the listener in music and speech. This information was felt to be necessary for a true evaluation of the CCIF method. RCA researchers reported the results of several experiments performed to clear up this point. The listeners' judgments were found to depend on the frequency band of the system. The report concluded that 2.5% is objectionable if the cutoff frequency is 15 kHz, while about five times that amount is still nonobjectionable when the cutoff

*CCIF: International Telephonic Consultative Committee.

is 3,750 Hz (Fig. 9.4) [5, 6]. Others report that according to the CCIF method of measuring nonlinear distortion 4 % is objectionable in the 400-Hz to 4-kHz region. If this is extended to RCA's 15-kHz limit, then at the latter frequency approximately 1 % should be found objectionable by the CCIF method.

9.3.1.6   Intermodulation distortion measurements by the CCIF method. An experimental test record for use with the CCIF method was made up. The CCIF or difference frequency method requires the measurement of the 440-Hz difference signal produced by the nonlinear reproduction of two equal recorded signal levels whose frequencies differ by 440 Hz. The distortion is equal to the magnitude of the difference signal divided by the magnitude of the recorded signals. This allows an evaluation of second-order distortion. Results of measurements by this method are shown in Fig. 9.5. The distortion

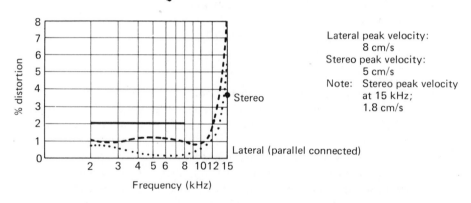

Figure 9.5   Distortion measurements by the CCIF method for a stereo phonograph cartridge.

percentages given are, of course, for the overall nonlinear distortion produced by the complete record reproduction system. The dimension of the burnishing facet of the recording stylus as well as its mechanical angle will also influence the high-frequency nonlinear distortion obtained upon playback. In addition, the radius of curvature greater than that of the reproducing stylus, the translation loss of the cartridge, and the mechanical stylus impedance of the cartridge will also affect high-frequency nonlinear distortion.

Figure 9.5 shows the distortion for a stereo cartridge with a vertical stylus force of 4 g, a bandwidth of 15 kHz, and an effective mass of 4.5 × 10^{-3} g. Beyond 11 kHz, the monophonic nonlinear distortion increases for this cartridge also. At 15,000 Hz this distortion is 8 %, which means that the distortion is $3\frac{1}{2}$ times lower than for the previous cartridge. Parallel channel operation for lateral playback results in a $2\frac{3}{4}$-fold reduction in distortion at 4 kHz. For stereo operation, the nonlinear distortion is only 3.5 % at 15 kHz; this represents a $4\frac{1}{2}$-fold reduction in distortion as compared to the previous cartridge. Correlation with other cartridges (magnetic and ceramic) using the same experimental test record shows a nonlinear distortion beyond 9 kHz of 2–10% for high-quality stereo cartridges and of 10–50% for poorer-quality cartridges.

9.3.1.7   Summary

1.   Nonlinear distortion exists in phonograph cartridges made from discs.

2.   The SMPTE method provides a measurement of nonlinearity at low frequencies for large excursions.

3.   The harmonic distortion method provides an analysis of distortion and nonlinearity to approximately 8 kHz.

4.   Only the CCIF method provides a measurement of nonlinear distortion near the upper frequency limit of the system.

5.   Turntable checks for speed, wow, and flutter and phonograph cartridge performance can be presently checked with the NAB (National Association of Broadcasters), monophonic-stereo 12-5-93 test record.

9.3.2   Amplifier Distortion*

9.3.2.1   Harmonics and nonlinear distortion.   Consider a perfectly pure 400-Hz sine-wave signal being applied to a recording amplifier. If the amplifier has no harmonic distortion, then the output waveform would be a magnified but otherwise wholly unchanged replica of the input [Fig. 9.6(a)]. If the outputs are as shown by the solid-line waveforms of Fig. 9.6(b) through 9.6(h), then harmonic distortion is present. The amplifier acts as though it were adding additional frequencies which, when added to the fundamental, result in the distorted waveforms shown in Fig. 9.6.

In a single-ended amplifier, the harmonics that are generated are mainly even harmonics. In push-pull amplifiers the odd harmonics are predominantly present since most of the even harmonics are cancelled out.

9.3.2.2   Amplifier total harmonic distortion.   The harmonic distortion factor of a signal is the ratio between the total rms values of the total rms value of the fundamental and all the harmonics. This factor can be expressed as a percentage (multiply by 100) and called a measure of the total harmonic distortion.

The percentage of total harmonic distortion (THD) is equal to the square root of the sum of the squares of all the harmonics divided by the square root of the sum of the squares of the fundamental and the harmonics, all multiplied by 100. See Fig. 9.15.

Assume we have a distorted waveform (fundamental plus harmonics) with an rms value of 50 V and we find that we have, in addition to the funda-

---

*Section 9.3.2 is based on Marc Saul, "Understanding Harmonic Distortion," *dB Magazine* (March 1976), pp. 24–26, 28 and 30–31.

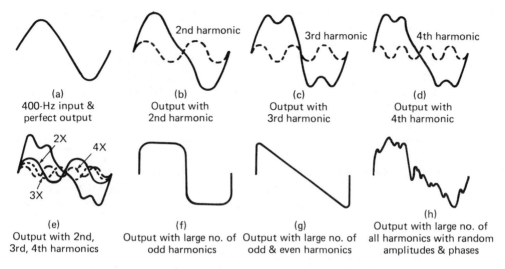

(a)
400-Hz input &
perfect output

(b)
Output with
2nd harmonic

(c)
Output with
3rd harmonic

(d)
Output with
4th harmonic

(e)
Output with 2nd,
3rd, 4th harmonics

(f)
Output with large no. of
odd harmonics

(g)
Output with large no. of
odd & even harmonics

(h)
Output with large no. of
all harmonics with random
amplitudes & phases

Figure 9.6   Waveforms with various numbers of harmonics.

mental signal, a second harmonic of 2 V and a third harmonic of 1.5 V. Our percentage of THD is the square root of 2^2 plus 1.5^2 or 2.5 divided by 50, all multiplied by 100, or 5 % THD.

Sometimes the distortion is weighted in proportion to the order of the harmonics. When this is done, the percentage of the individual harmonics is multiplied by a weighting factor that increases as the order of the harmonic increases.

With just about every system of amplification other than class B, the percentage of total harmonic distortion decreases as the power output level is reduced. In addition, as the output power level is reduced the percentages of the higher-order harmonics decrease more rapidly than those of the lower-order harmonics. This means that the THD usually decreases as the power output is reduced. However, in some transistor amplifiers, when you go down to very low output powers, THD may actually begin to rise again slightly.

When negative feedback is used, all the harmonics are reduced in the same proportion. This does not affect their relative importance except when the overload point is reached.

9.3.2.3   How much system distortion are we stuck with?   An important question is just how much distortion we can perceive or tolerate in an audio system. A number of tests were conducted by RCA researchers [5] using a single-ended, low-power (3-W) amplifier. The tests were conducted in a typical living room environment with a noise level of about 25 dB. A limited number of critical observers were used to rate the intentionally introduced distortion from objectionable through tolerable to perceptible.

The results of these tests are as follows: For an amplifier high-end cutoff of 7,500 Hz, objectionable distortion occurred at 4–4.8% THD for music and at 6.4–6.8% THD for speech. Tolerable distortion occurred at 3.2–4.4% for music and at 4–4.8% THD for speech. Perceptible distortion occurred at 0.95% THD for music and at 1.15–1.2% THD for speech.

Next, the high-end cutoff was extended to 15,000 Hz. Under these conditions, objectionable distortion was 2.0–2.5% for music and 3.0–4.4% for speech. Tolerable distortion occurred at 1.35–1.8% for music and at 1.9–2.8% for speech. Perceptible distortion occurred at 0.7–0.75% with music and at 0.9% with speech.

9.3.2.4   Measuring amplifier total harmonic distortion.   A simple method of measuring total harmonic distortion is to use a distortion analyzer, shown in Fig. 9.7. A low-distortion audio generator is applied to the input of the

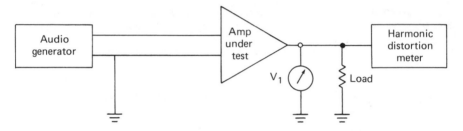

Figure 9.7   Test setup used to measure total harmonic distortion of amplifiers.

audio unit under test with its output properly terminated. The output of the amplifier is monitored by a voltmeter V_1 and is fed to a harmonic distortion analyzer. The meter, or analyzer, will read the percentage of THD directly.

The harmonic distortion meter contains a selective audio voltage amplifier, with adjustable attenuation, whose output is connected internally to a high-impedance voltmeter circuit. The purpose of the selective amplifier is to suppress or null out the fundamental frequency, resulting in the analyzer measuring the remaining harmonics.

The usual method of obtaining this selectivity is by the use of a tunable band rejection filter network, shown in Fig. 9.8(a). When the notch is adjusted to the fundamental frequency of the audio generator, the fundamental frequency is effectively removed or suppressed by 60–80 dB. The meter in the harmonic distortion analyzer now has applied to it all the other components of the waveform. These are mainly harmonics.

With the amplifier distortion being a function of amplifier power output, the input signal is increased such that the amplifier output is at rated power. The amplifier power output is obtained by monitoring the output voltage,

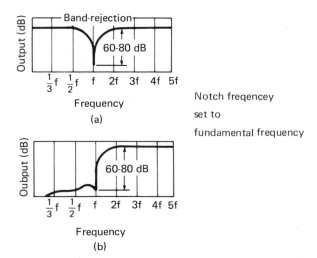

Notch freqencey

set to

fundamental frequency

Figure 9.8 The two types of response curves used in harmonic distortion analyzers.

V_1, with the power output calculated by

$$P_o = \frac{V^2_{o(\text{rms})}}{R_L} \qquad (9.1)$$

The amplifier THD is measured as follows: The analyzer's selective amplifier is initially bypassed entirely, and a meter reading is taken of the output of the amplifier under test. This reading is the value of the fundamental frequency plus the distorting harmonics. The meter is now adjusted for full-scale (100 %) reading. The selective amplifier is switched in and the notch frequency is adjusted (fundamental suppression) for a null. The resulting meter reading is a direct indication of the percentage of total harmonic distortion.

Some harmonic distortion meters employ sharp cutoff high-pass filters to eliminate the fundamental frequency [Fig. 9.8(b)]. With such a curve, not only is the fundamental frequency removed but the hum and noise below the fundamental are also effectively eliminated. As such filters are not usually adjustable, these instruments may use a half dozen or so filters with different cutoff frequencies in order to permit measurements to be made at various fundamental frequencies. In some cases, two such filters are used, one cutting off at 400 Hz and the other at 1,000 Hz. This permits THD measurements to be made at these two fundamental frequencies.

In general, as the output power or voltage is increased, so is the amount of distortion. Usually, the increase is smooth and gradual up to the overload point, where there is a sudden increase in distortion. Amplifiers should be rated at a power or voltage just below this overload point, while the THD is still only a small value, say 1 %.

In Fig. 9.9 we see an amplifier's percentage of THD (at some midfrequency) plotted against its output power. In this case, the amplifier has a THD of less than 1 % below output powers of 50 W. At 50W, the THD is just 1 %. Above this power, the THD rises quickly to a value of 5 % at 60 W. This amplifier would then be rated at 50 W with 1 % THD.

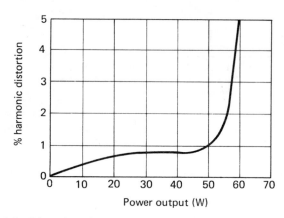

Figure 9.9 Distortion characteristic of a typical hi-fi power amplifier.

A high-quality 50-W amplifier whose THD has been measured over the entire audio range may have the following specification: "Total harmonic distortion at 1 % or below from 20 to 20,000 Hz within 1 dB of 50 W." Such an amplifier would have no trouble at all in producing up to a full 50 W of output at 1 % or less of THD over most of the audio range. At the very extremes of the range, however, it would still be producing 1 % or less distortion at powers up to 40 W (which is 1 dB below 50 W).

9.3.2.5   Using a scope to observe amplifier distortion.   An oscilloscope may also be used to observe ha·monic distortion in an amplifier or other piece of audio equipment. As a rule, it is difficult to see distortion much less than 3–5 % on a 5-in. cathode ray tube. In some cases, though, especially with higher-order harmonics, some 2–3 % distortion can be observed. It is important that the scope used have vertical and horizontal amplifiers with similar or equal frequency responses and phase characteristics.

To check distortion with a scope, use the setup shown in Fig. 9.10 The audio generator is fed to the input of the amplifier under test and is also applied to the horizontal input terminals of the "scope," with its horizontal sweep frequency turned off. The output of the amplifier, while being moni-

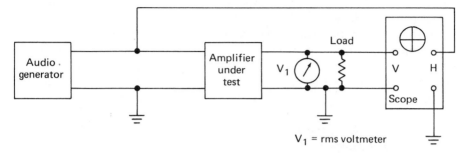

Figure 9.10 Test setup for oscilloscope observation of harmonic distortion.

tored by a voltmeter, V_1, is properly terminated with a load resistor and is applied to the vertical input terminals of the oscilloscope. The frequency of the audio generator is usually set at some convenient middle frequency, such as 400 or 1,000 Hz, or at any frequency in which THD is to be measured. The voltmeter, V_1, monitors the output V_{rms} voltage, from which the output power is determined.

The waveform seen on the "scope" will be a perfectly straight diagonal line, for zero distortion, assuming that the scope's gain controls are adjusted so that equal voltages are applied to the deflecting plates of the cathode ray tube [Fig. 9.11(a)]. This straight line is actually the transfer characteristic of

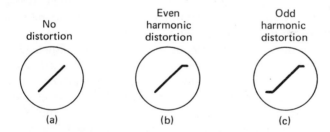

Figure 9.11 Distortion patterns on an oscilloscope being used to observe harmonic distortion.

the amplifier under test., With the amplifier having harmonic distortion, the straight line will begin to show some curvature at either one or both ends, or it may show some curvature somewhere along the length of the trace. If the curvature is at one end [Fig. 9.11(b)], we are seeing the results of even harmonic distortion, and if the curvature is at both ends [Fig. 9.11(c)], we are seeing the results of odd harmonic distortion.

A drawback of this technique is that we cannot readily determine the actual percentage of harmonic distortion. About all that can be done is to determine whether or not distortion is present. The more the nonlinearity or distortion, the greater will be the curvature of the scope trace.

A measurement of total harmonic distortion, although it does not tell the entire story about a unit's characteristics, is one of the most useful performance specifications that can be measured. A third method for measuring harmonic distortion is by using a spectrum analyzer, discussed in Section 9.4.

9.3.3   Introduction to Loudspeakers—Woofers and Tweeters

Loudspeakers are the last electromechanical link in a high-fidelity system. Such devices may be termed transducers or sensors and will be discussed in Section 11.9 in detail. As previously shown in Fig. 9.1, the woofer is the low-frequency speaker which encompasses a frequency range from 30 to 5,000 Hz, with the high-frequency loudspeaker or tweeter having a frequency range of from 5,000 to 20,000 Hz. The crossover network in the loudspeaker system consists of one or more capacitors-inductors connected in appropriate fashion as a high-pass filter, directing the low frequencies to the woofer and the high frequencies to the low-power tweeter.

Any major resonances in the loudspeaker system appear as a rise reaching a peak and then a drop in the frequency response. For the low-frequency speaker or woofer, the natural resonance can be as low as 30 Hz, while the tweeter can have a natural resonance of 15,000 Hz. These resonances govern to an extent the frequency range of the loudspeaker system. We should further note that the speakers are mounted in a cabinet for the sounds to be properly dispersed in the medium of air. To determine woofer performance, total harmonic distortion measurements are adequate. Such techniques as described in Section 9.3.2.2 are appropriate for total harmonic distortion measurements of woofers with the resistive load replaced by the woofer under test. A total harmonic distortion figure of 6% is tolerable. A microphone with a flat frequency response (20-20 kHz) is required to convert the loudspeaker output (an acoustical signal) to an electrical signal for measurement with a distortion sensing instrument.

It is important to realize that every reproducing system can be sophisticated, consisting of as many as *four* woofers and *four* tweeters. A three-way loudspeaker system is found in many designs with a midfrequency speaker filling in the frequency spectrum roughly between 500 and 5,000 Hz. This speaker system reduces intermodulation products and minimizes the modulation of the higher frequency by the lower, maintaining a broad directivity pattern. As the frequency is increased the low-frequency speaker (the woofer) becomes directive, requiring a smaller midrange loudspeaker to broaden the directivity pattern.

9.3.4 High-Frequency Loudspeakers

9.3.4.1 Types and causes of distortion.* High-frequency loudspeakers or tweeters may have the following types of distortion:

1. *Nonlinear distortion,* resulting from (1) deviation from Hooke's law (nonlinear displacement of suspension versus applied force), (2) nonuniform flux density of the magnetic field in which the voice coil moves, (3) very high pressure amplitudes at the throat of the horn types, and (4) mechanical defects such as rubbing of the voice coil in direct radiator and horn types, diaphragm "hitting" of the phasing plug in dynamic horn types, or excessive diaphragm displacement relative to spacing in electrostatic types.

2. *Frequency distortion,* a linear distortion whereby a change in waveform is brought about by a change in the relative magnitudes of the different frequency components of a complex wave. Frequency distortion at high frequencies may be severe inasmuch as the diaphragm can assume a different mode of vibration: Instead of oscillating as a whole, it breaks up its vibration into a standing wave pattern.

3. *Transient distortion,* also a linear distortion, involves the degree to which a tweeter fails to respond accurately to abrupt changes in the input signal level. This type of distortion is usually due to improper damping of the moving parts.

4. *Frequency modulation distortion or flutter:* A single cone loudspeaker for low and high frequencies will generate distortion even though the motion of the diaphragm is linear. The cone excursions at low frequencies are large, with a high-frequency source causing the cone to move back and forth (at the rate and amplitude of low frequencies) such that the high frequency becomes frequency-modulated due to the Doppler effect.

The distortion of an 8-in. speaker in a flat baffle radiating $\frac{1}{20}$ W at 50 Hz simultaneously with a 15-kHz signal would be approximately 37%. This type of distortion can be minimized by the use of separate high- and low-frequency speakers.

9.3.4.2 Distortion measurement techniques

1. *Intermodulation Measurements:* If the original signal consists of one or more frequency components, the resulting signal will also have terms corresponding to the sum and difference of integral multiples of the input frequencies. This is called intermodulation distortion. Thus the output will contain frequencies which are not harmonically related to the input signals.

There are two intermodulation methods used for the evaluation of nonlinear distortion. The SMPTE method evaluates the nonlinear distortion for a

---

*Sections 9.3.4.1 through 9.3.4.3 including Figs. 9.12 through 9.15 are based in part on P. Kantrowitz, "Distortion Measurements of High-Frequency Loudspeakers," *Journal of the Audio Engineering Society,* Vol. 10, No. 4 (Oct. 1962), pp. 310–317.

large-amplitude low-frequency signal in the presence of a high-frequency signal amplitude that is one-fourth of the low-frequency signal amplitude. The distortion terms are determined from a measurement of the sum and difference frequencies around the high-frequency signal. The other intermodulation method is the CCIF or difference-frequency method, which involves the measurement of a difference-frequency signal produced by the nonlinearity of the tweeter resulting from two equal amplitude signals. Here the distortion is expressed as the ratio of the magnitude of the difference-frequency signal to the sum of the two input signals. In the analysis of high frequencies near the upper frequency limit of the tweeter, this method is the only one which gives reliable results.

2. *Harmonic Measurements:* If a pure sine wave is applied to a nonlinear system, the resultant waveform will not be a pure sine wave but will contain frequencies that are harmonics or integral multiples of the original frequency.

The harmonic distortion method is useful only in the detection of distortion at frequencies at most one-half of the upper cutoff frequency. Where cubic distortion is predominant, the measurement of total harmonic distortion is inadequate for providing information for frequencies greater than 7 kHz for a 20 kHz system.

9.3.4.3 Distortion measurements for tweeters.   The microphone required for such measurements should have a relatively smooth response to 20 kHz. In addition, the test room used for the distortion measurements and subjective listening tests should have acoustic characteristics representing that of an average living room.

The response of each speaker is measured using a random noise source. The noise method of testing loudspeakers was used here rather than a sinusoidal signal technique because noise more nearly approximates the time-amplitude characteristics of program material; the standing waves in the testing room were also greatly reduced.

A block diagram of the equipment used for the loudspeaker response measurements is shown in Fig. 9.12. A filter having a slope of -3 dB/octave from 20 Hz to 20 kHz converts the white noise from constant energy per cycle to constant energy per octave, which has been called *pink noise*. The pink noise is applied to the loudspeaker, and the response is recorded after passing through a narrow-band (8% wide) filter. The sound pressure level (SPL) is 90 dB re 0.0002 μbar. This is equivalent to the level of a symphony orchestra at 20 ft and would be the maximum to which a normal listener would be subjected in an average living room.

A block diagram of the equipment used to make the intermodulation distortion measurement is shown in Fig. 9.13; Fig. 9.14 is a block diagram of harmonic distortion measurements on a loudspeaker system.

Refer to Fig. 9.15 for the frequency spectrum of the input and distortion

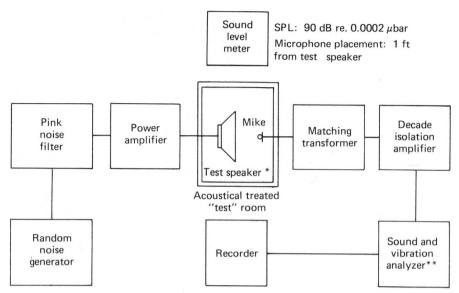

Figure 9.12   Block diagram of equipment used to make response measurements.

components and distortion terms for CCIF intermodulation and harmonic distortion terms.

It should be noted here that the above measurements can be performed with a spectrum analyzer. See Section 9.4.

When the total CCIF distortion is less than approximately 3%, smoothness of response and extent of range combined with directional characteristics can play a more important role than nonlinear distortion in a subjective listening evaluation of high-frequency loudspeakers.

Consideration should be given to the disc recording-reproducing process, since this process can generate more distortion than a quality loudspeaker system.

9.3.4.4   Units and conversion

1.   The sound pressure level (SPL) is sound pressure related logarithmically to a reference level of pressure (P_o) which by convention is 0.0002 dyn/cm² or 2×10^{-4} μbars or 2×10^{-5} newton/m².

2.   The decibel (dB) is a comparison of two levels, in particular power or intensity of sounds. If one sound is 100 times as powerful, it is said to have a level of 20 dB with respect to the first sound. This is the case since Weber-Fechner found that the stimulus of the intensity signal followed a logarithm ratio, which is another way of expressing powers of 10.

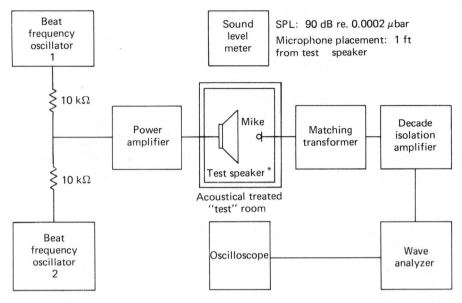

*4 μF in series with test speaker unless otherwise specified.

Figure 9.13 Block diagram of equipment used to make intermodulation distortion measurements by the CCIF method.

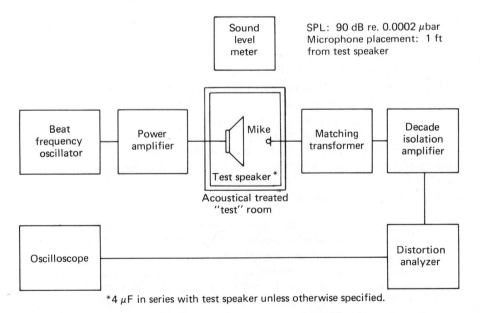

*4 μF in series with test speaker unless otherwise specified.

Figure 9.14 Block diagram used to make total harmonic distortion measurements.

244

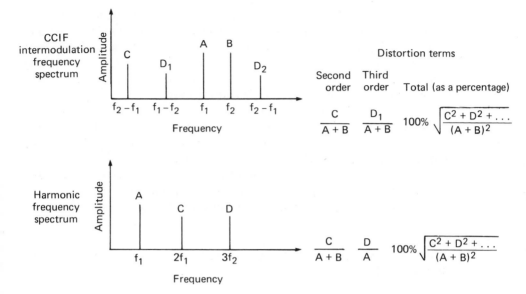

Figure 9.15 Frequency spectrum of input and distortion components and distortion terms for CCIF intermodulation and harmonic distortion terms.

3.   The microbar (μbar) is a unit of pressure used in acoustics. One microbar is equal to 1 dyn/cm^2.

4.   An octave is the interval between two sounds having a basic frequency ratio of 2.

5.   White noise is a noise spectrum whose noise power in a frequency band, ΔF, is the same ΔF whether the frequency is 20 or 20,000 Hz. White noise need not be random.

9.4   SPECTRUM ANALYZERS

9.4.1   Introduction to Spectrum Analysis

The wave analyzer is a manually operated instrument, with the ability to measure each harmonic individually. This is achieved by beating the input signal spectrum with an internal oscillator and then filtering the mixer products to obtain a signal voltage that is proportional to the harmonic voltage. The signal voltage is then measured on an ac voltmeter calibrated to read the harmonic rms value. The wave analyzer scanning range equals the internal oscillator tuning range, and the wave analyzer window equals the filter amplifier bandwidth.

Today, after many years of electronic development, the spectrum analyzer [7, 8], which performs like a wave analyzer except that the internal oscillator is swept electronically through the scanning range and the final indication appears on an oscilloscope, is widely used in audio work. At any instant in time, the spectrum analyzer is looking at some part of the input spectrum through a narrow window. If a signal is present, a vertical line appears on the oscilloscope. The vertical line position along the base line is an indication of the frequency component involved.

Early spectrum analyzers were nothing more than indicators, giving a representation of frequency and amplitude. The introduction of the amplitude calibrated spectrum analyzer several years ago triggered the development of the instrument into the measurement tool it is today.

We can say that the spectrum analyzer is a tuned receiver with selectable frequency ranges and spans, intermediate frequency bandwidth, and a linear or log detector—all coupled to the cathode-type display. Spectrum analyzers are presently available from the low audio-frequency range to microwaves.

A simplified block diagram of a modern swept front-end three-knob spectrum analyzer is shown in Fig. 9.16.

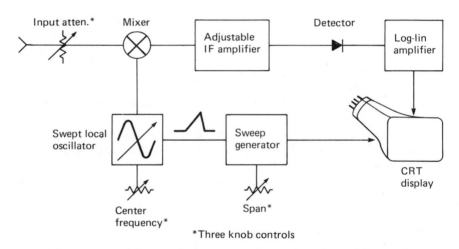

Figure 9.16 Simplified block diagram of a modern swept front-end three-knob spectrum analyzer. Courtesy of Tektronix Inc., Beaverton, Oregon.

Refer to Figs. 9.17(a) and (b) for photographs of a spectrum analyzer plug-in model 7L12, which plugs into a 7613 oscilloscope. The analyzer frequency response is from 0.1 to 1800 MHz.

A simplified block diagram of a tracking generator is given in Fig. 9.18. The tracking generator is a signal source that follows the spectrum analyzer tuning. This instrument combination is ideal for making frequency response measurements. The tracking generator is locked into the spectrum analyzer.

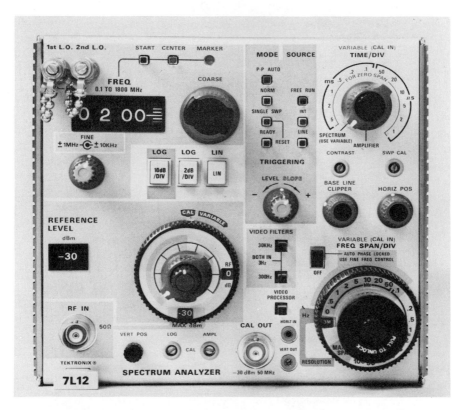

Figure 9.17   (a) Model 7L12 plug-in spectrum analyzer. Courtesy of Tektronix, Inc., Beaverton, Oregon.

9.4.2   Real-Time Analysis Using Spectrum Analyzers

The problem of speed is attacked by *real-time analysis*, implying the use of a large number of fixed-frequency filters (or resolution elements) rather than the use of one sweeping filter.

The basic meaning of *real time* is "very fast." Much more data can be processed with the real-time analyzer (RTA) than with a non-real-time instrument such as the traditional sweeping analyzer. This difference is so enormous that some experiments would never be attempted with a non-real-time unit, because of the months it might take for an analysis. Real-time analyzers [7, 8] simultaneously display the amplitude of all signals in the range of the analyzer—hence the name "real time."

Typically, a 500-line (i.e., 500 filter elements) RTA processes 500 times as fast as a sweeping analyzer of equivalent bandwidth and range. In terms of time, this means that 100 analyses and plots, which can be completed in less than 4 h with the RTA, would take many months with a sweeping unit.

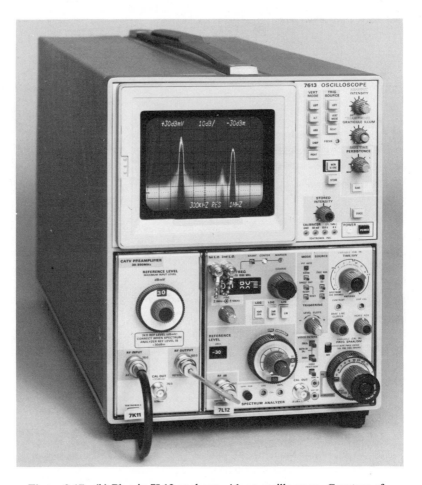

Figure 9.17 (b) Plug-in 7L12 analyzer with an oscilloscope. Courtesy of Tektronix Inc., Beaverton, Oregon.

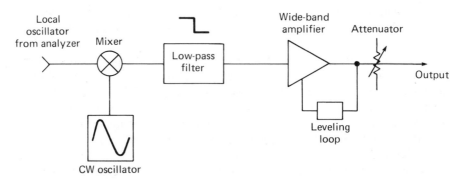

Figure 9.18 Simplified block diagram of a tracking generator. Courtesy of Tektronix Inc., Beaverton, Oregon.

Frequently, real-time analysis is synonymous with *on-line* analysis. For example, a test can be monitored as it is being run by watching particular frequencies that might warn of impending catastrophe, or a manufacturing test that requires an immediate decision can be made on a production line. More commonly, however, *on-line* spectrum analysis is used to preview data being taped, to verify their quality, and to anticipate the value of the frequency data to be plotted later in the laboratory.

Another meaning of "real time" is the ability of the RTA to sweep through the spectrum so rapidly that the display will not appear to flicker on a standard oscilloscope. Twenty sweeps per second (50 ms/sweep) are sufficient.

9.4.3 High-Fidelity Measurements with the Audio Tracking Spectrum Analyzer*

Maintaining a modern high-fidelity-stereo system today requires much more than a "trained ear." The high specifications of receivers and amplifiers can only be maintained by performing some of the standard measurements, such as

1. Power output.
2. Harmonic distortion.
3. Intermodulation distortion
4. Frequency response.
5. Signal-to-noise measurements.
6. Distortion vs. output.
7. Power bandwidth.
8. Damping factor.
9. Square-wave response.
10. Transient intermodulation distortion.

Unfortunately, because of the available test equipment and lengthy procedures that are required to "spec" a new or repaired amplifier, it usually doesn't get done.

In this section we shall describe an alternative test device and procedures that permit rapid, easy to understand, "spec'ing" and troubleshooting. The device, a low-frequency spectrum analyzer, can perform standard measurements and may be used effectively for expanding the standard tests or for special measurements such as the CCIF distortion or Bell Telephone multi-tone audio tests.

---

*Excerpted from *The Tektronix Cookbook of Standard Audio Tests*, copyright 1975, Tektronix, Inc., Beaverton, Oregon. Reproduced by courtesy of Tektronix, Inc.

Frequency response and intermodulation distortion measurements will now be discussed using a low-frequency spectrum analyzer (Tektronix 5L4N or equivalent). The two measurement parameters give first-hand insight into the system behavior.

9.4.4 Frequency Response

Frequency response is a measure of the amplifier's ability to pass a wide range of frequencies in the audio spectrum. Ideally, we strive to achieve a flat response; that is, all frequencies would pass through an amplifier with equal amplification. A hi-fi amplifier may have controls to modify the response. These may include tone controls (bass and treble), rumble and hum filters (low-frequency roll-off), scratch filters (high-frequency roll-off), and a variety of tailoring devices such as the RIAA, FM deemphasis, and tape head equalization filters. The frequency response test should provide response information of the amplifier in the flat position and should also represent the limits and interaction of the tone controls and filters.

Response of a modern hi-fi system is generally measured from below 20 Hz to well beyond the 15-kHz audible limit. It is measured in decibels of deviation across the audio spectrum.

The analyzer is ideally suited to frequency response testing since it has a self-contained tracking generator and a log sweep 20-Hz to 20-kHz mode. An amplifier can be swept under a variety of different conditions in a matter of seconds, eliminating the need for tedious measurements and point-to-point plots. Multiple traces of conditions can be built up either on film or on a storage oscilloscope to obtain one picture of the complete response performance of an audio device.

The rated frequency response [9] is the frequency range over which the amplitude response does not vary more than plus or minus 3 dB from the amplitude at 1,000 Hz.

A load matching chart and a frequency response equipment setup are shown in Figs. 9.19 and 9.20, respectively.

Frequency response measurements made at 10%, 50%, and 75% power and tone control ranges are shown in Figs. 9.21 and 9.22, respectively.

9.4.5 Intermodulation Distortion—General Discussion

Intermodulation distortion is determined by feeding two or more pure tones into an amplifier and measuring the amount one tone is transferred (cross-modulated) onto the other.

Two commonly used intermodulation tests utilizing a spectrum analyzer are described in Fig. 9.15 and by the chart in Fig. 9.23. See also Section 9.3.4.2.

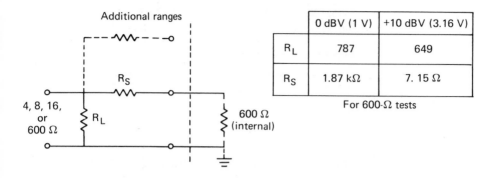

Additional ranges

	0 dBV (1 V)	+10 dBV (3.16 V)
R_L	787	649
R_S	1.87 kΩ	7. 15 Ω

For 600-Ω tests

	R_S			
R_L	1 W	10 W	50 W	100 W
4 Ω	*4.87 kM	15.4 kΩ	34.8 kΩ	48.7 kΩ
8 Ω	6.98 kΩ	22.1 kΩ	48.7 kΩ	69.8 kΩ
16 Ω	9.76 kΩ	30.9 kΩ	69.8 kΩ	97.6 kΩ

$$R_s = \frac{\sqrt{R_{load} \times R_{full\ screen} \times R_{analyzer\ impedance}}}{V_{full\ screen}}$$

Full screen = 0.245 V into 600 Ω
*All resistors 1% tolerance

Figure 9.19 Load matching chart.

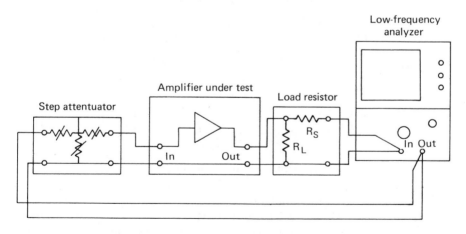

Figure 9.20 Frequency response equipment setup.

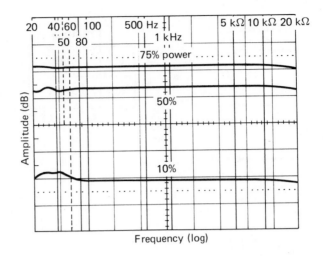

Figure 9.21   Frequency response at 10%, 50%, and 75% power.

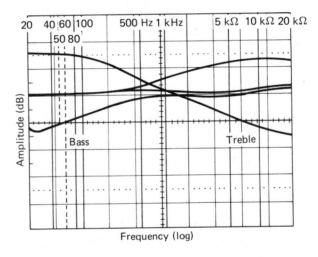

Figure 9.22   Tone control ranges.

Measurement of intermodulation distortion [10] has always generated a lot of controversy. All involved agree that the various measurements produce numbers that relate to performance of audio equipment; however, everyone has his/her own idea about what frequencies to use, how many, the levels, etc.

The low-frequency spectrum analyzer can handle all of the different known methods. The procedure presented below uses the Society of Motion Picture and Television Engineers (SMPTE) modulation method using two

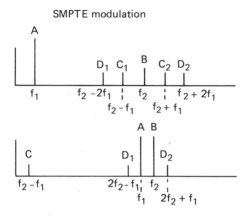

SMPTE modulation

CCIF difference intermodulation

Distortion		Commonly used frequencies
2nd	3rd	
$\dfrac{C_1 + C_2}{B}$	$\dfrac{D_1 + D_2}{B}$	60 Hz — 6,000 Hz 60 Hz — 7,000 Hz (1 HFM)
$\dfrac{C}{A + B}$	$\dfrac{D_1}{A + B}$	6,000 Hz — 7,000 Hz 1,200 Hz — 1,400 Hz

Figure 9.23   IM methods of measurement.

tones. The specific tones and ratios are recommended by the Institute of High Fidelity Manufacturers (IHFM).

The equipment setup used to make intermodulation distortion measurements using the low-frequency spectrum analyzer is shown in Fig. 9.24. A SMPTE modulation method generator output is shown in Fig. 9.25, with the resulting SMPTE modulation of measurement illustrated in Fig. 9.26. A CCIF difference method of measurement is given in Fig. 9.27.

9.4.6   Transient Intermodulation Distortion

Transient intermodulation distortion (TIM) [11, 12, 13] is distortion in amplifiers that occurs principally during loud, high-frequency passages. Most music contains some material that can cause TIM distortion. Amplifiers with large amounts of negative feedback are prone to TIM distortion because the amplifier loop, if improperly designed, requires too much time to respond to rapid transients.

Ever since the introduction of the transistor power amplifier, the *transistor sound* has been discussed. Even though in many cases a transistor amplifier tested better in terms of distortion than a tube counterpart, during a listening test the tube unit would unmistakably perform better. TIM distortion is one explanation of these discrepancies. Transistor amplifiers test "excellent" using steady-state harmonic and intermodulation tests. However, music material generates amplifier distortion because of its transient nature.

A popular explanation of the source of TIM is that the transient reaches or exceeds the slew rate of the amplifier, causing an instant severe intermodulation condition until the negative feedback signal catches up with and corrects the distortion.

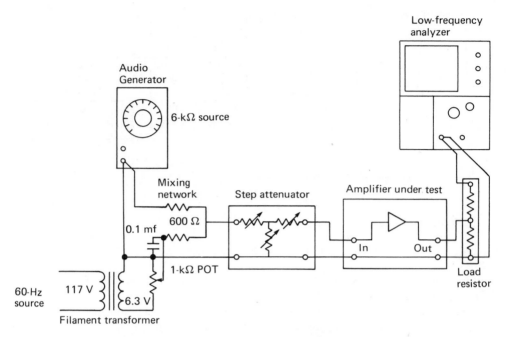

Figure 9.24   Equipment setup for intermodulation measurement.

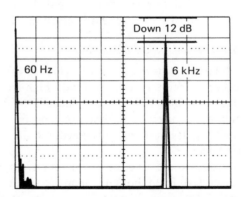

Figure 9.25   SMPTE modulation method generator output.

No measurement standards exist to date. However, a square wave with a high-frequency sine wave has been used to observe this distortion.

Presented below is a technique that used a 6-kHz sine wave mixed with a 500-Hz square wave to demonstrate TIM distortion. However, no single number results to adequately indicate the amplifier's performance.

See Fig. 9.28 for an equipment setup for TIM demonstration using a low-frequency spectrum analyzer.

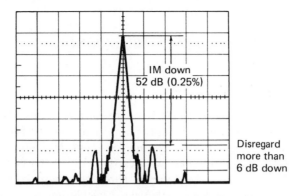

Figure 9.26 SMPTE modulation method of measurement.

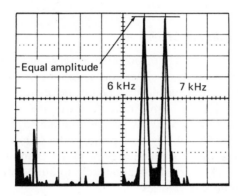

Figure 9.27 CCIF difference method of measurement.

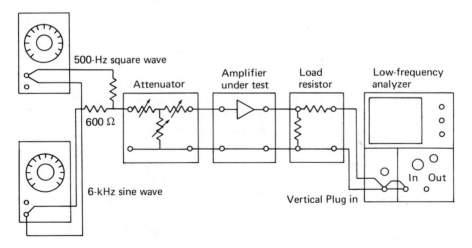

Figure 9.28 Equipment setup for TIM demonstration.

255

9.5   OTHER AUDIO-FREQUENCY MEASURING
INSTRUMENTS*

9.5.1   Sound Level Meters

The purpose of a sound level meter is to measure in terms of a standard reference level the acoustic power at any point in an acoustic field.

The acoustic pressure reference chosen for sound level studies is 0.0002 microbars, the intensity equivalent of which is 10^{-12} W/m². This level is designed as 0 dB, and an energy vs. frequency distribution is required to produce a sensation of equal loudness. Filter networks or weighting filters are incorporated in modern sound level meters to make their response correspond to a corrected energy vs. frequency curve.

9.5.2   Volume Indicator Meters

These meters apply to a variety of portable meters commonly used for measuring audio power in terms of either watts, volume units (vu), or decibels. These include audio power output meters, decibel meters, vu meters, etc.

Some laboratory-type portable output power meters have a full-scale power range from 10 mW to 100 W, corresponding to a range of $+10$ to 50 dB above 1 mW. The ability to operate with terminating impedances from 2.5 to 20,000 Ω, with a frequency error of less than 1 dB and from 20 to 15,000 Hz, makes such instruments applicable to almost any type of power measurement in audio laboratories.

Laboratory-type audio indicators should not be confused with commonly used ac voltmeters, which may carry a decibel scale for interpreting ac volts across a standard 500- to 600-Ω line in terms of decibels referred to 1 mW. Any ac voltmeter may be used in such a way by merely applying the formula

$$dB = 10 \log_{10} \frac{E^2}{0.001z} \tag{9.2}$$

where E is the measured ac voltage and z is the line impedance.

The vu is a standard term for designating absolute levels of audio power above or below a zero-reference level of 1 mW in 600 Ω. The vu is numerically equal to the number of decibels above or below 1 mW. The volume indicator meter is a device used to monitor recorded or broadcast signals in order that the dynamic range of the transmitted signal will not exceed the modulation capabilities of the transmitter or recording system. In a broadcasting station, the vu meter will read 100% modulation at 0 vu.

*Section 9.5 is based on material from *Tel-Communication Engineers Instrument Manual*, Caldwell-Clements, Inc., New York, 1947.

256

9.5.3   Present Standard Power and Voltage Unit Terms

1.   The amplifier power level of 0 dBm refers to the audio power level of 1 mW. 1 dBW indicates an audio power level output of 1 W. A power output level of 1 dBW is also equivalent to a power level of 30 dBm.

2.   The voltage level L_U in decibels is 20 times the logarithmic to the base 10 of the ratio of the voltage under consideration U in volts to the reference voltage U_{ref} in volts, or, mathematically,

$$L_U \text{ re. } U_{ref} = 20 \log \frac{U}{U_{ref}} \qquad (9.3)$$

9.6   RATINGS OF AUDIO ELECTRONIC DEVICES

9.6.1   Cartridges

Cartridges are rated in volts or in fractions of volts, with frequency response also being an important rating. Modern monophonic (single-channel) cartridges can have a frequency response as high as 20,000 Hz $\pm$ 1,000 Hz when tested with a CBS-STR-100 test record.

There is presently a CD-4 (discrete four-channel record) made by Victor Company of America which can measure stereo cartridges to 45,000 Hz. The CD-4 test record contains two signals: On one side there is front and rear information like a regular stereo disc. The other channel is an ultrasonic carrier which conveys a differential signal between the front and rear information in frequency modulation. The resulting frequency spectrum limit is 45,000 Hz, so a stereo cartridge must be able to produce frequencies beyond 20,000 Hz. The output recorded on the CD-4 disc must be greater than 1 mV. The CBS-STR-100 test record can perform frequency measurement for stereo cartridges up to 50 kHz. There are presently stereo cartridges that have a frequency response from 30,000 to 40,000 Hz. The crosstalk (the separation between the two signal channels) for a CD-4 test record is approximately 15 dB at 30,000 Hz. The separation is what appears on the second signal when the first signal is played. Tone arms can affect the low-frequency response, with the present arms having a resonance of 7 Hz. Great strides have been made in phonograph cartridge frequency response performance, separation, and low tone-arm resonance, which should also be specified in stereo phonograph cartridges.

9.6.2   Microphones

Microphones [14] are rated at -50 to -120 dB re. 1 V/μbar. A typical microphone may have a sensitivity of -60 dB re. 1 V/μbar or 1 mV/μbar.

9.6.3 Preamplifiers and Power Amplifiers

Preamplifiers are rated in fractions of volts such as millivolts. Preamplifiers can also be rated in voltage gain, usually given in decibels. Power amplifiers are usually rated in peak power. Power rating is *not* on a continuous wave basis. The peak power can be given at a specified duty cycle (i.e., 50 W peak, 10% duty cycle). Amplifiers can have a power rating of say 50 W of average sine-wave power (14.2 V rms at 0.1% third-harmonic distortion) measured across an 8-Ω resistive load, which represents the speaker impedance. Continuous power output is that output measured with a sinusoidal signal for at least 30 s. Momentary power output (music power) is defined as that obtainable with a sine wave signal during a time interval so short that significant voltages have not changed appreciably from their no-signal values.

9.6.4 Loudspeakers

Loudspeakers are rated in sound pressure levels under various environmental conditions and placement in the listening room. The power amplifier and loudspeaker ratings may be given together. The electrical impedance of a loudspeaker varies from 4 to 16 Ω.

9.7 SUMMARY

Audio systems have been evaluated from the standpoint of distortion and parameters which govern their performance. Also, the units, ratings, and meters used in audio systems have been explained. These concepts seem simple but in actuality are quite complex. However, with the spectrum analyzer, audio measurements are performed more easily and quickly.

REVIEW QUESTIONS

1. What is meant by a high-fidelity system? Show a block diagram.
2. Describe how a cartridge harmonic distortion measurement is made.
3. Describe how cartridge intermodulation distortions are made.
4. Why are listening tests important in the evaluation of a phonograph cartridge?
5. Describe how amplifier harmonic distortion is made.
6. Describe how amplifier intermodulation distortion measurements are made.
7. What is meant by harmonics?
8. What is the value of using square-wave testing of amplifiers?

9. List the distortion terms found in high-frequency loudspeakers.
10. Describe how high-frequency loudspeaker harmonic distortion measurement is made.
11. Describe how high-frequency loudspeaker intermodulation distortion measurements are made.
12. What is the SPL?
13. What is the dB in relation to the power ratio? The voltage ratio?
14. What does 60 dB re. 1 V mean?
15. What is a microbar?
16. What is white noise?
17. What is a spectrum analyzer?
18. Why is a spectrum analyzer used in audio work?
19. What is meant by real-time analysis?
20. How are frequency response and intermodulation distortion measurements made with an audio spectrum analyzer?

REFERENCES

1. JAMES V. WHITE, "A Linear Theory of Phonograph Playback," *JAES*, *19*, No. 2 (Feb. 1971), pp. 94–100.
2. F. E. TERMAN and J. M. PETTIT, *Electronic Measurements*. McGraw-Hill, New York, 1952, p. 341.
3. A. PETERSON, "Intermodulation Distortion: Its measurement and Evaluation," *1957 IRE National Convention Record, 5,* Part 7 (1957), pp. 51-58.
4. H. E. ROYS, "Analysis by the two-frequency Intermodulation Method of Tracing Distortion Encountered in Phonograph Reproduction," *RCA Review*, Vol 10, No. 2 (June 1949), p. 254.
5. H. F. OLSON, "Stereophonic Sound Reproduction in the Home," *J. Audio Eng. Soc., 6* (1958), p. 82.
6. W. A. VAN BERGEYK, J. R. PIERCE, and E. E. DAVID, Jr., *Waves and the Ear*. Doubleday, New York, 1960, p. 202.
7. Spectrum Analyzer Basics, Application Note 150, *Spectrum Analyzer Basics*. Hewlett-Packard, April 1974.
8. *Measurement and Control, Handbook and Buyers Guide, 1974–1975*. Copyright 1974 Measurement and Data Corp., Pittsburgh.
9. EIA Standard Methods of Measurement for Audio Amplifiers Used in Home Equipment, *RS-234-C*. Electronic Industry Association Standard, 2001 I St. N. W., Washington D. C. 20006, 1971.

10. GERALD STANLEY and DAVID MCLAUGHLIN, "Intermodulation Distortion: a Powerful Tool for Evaluating Modern Audio Amplifiers," Audio, Vol 56, (Feb. 1972), pp. 36, 38, 40, 44, 45.

11. MATTI OTALA, "Transient Distortion in Transistorized Audio Power Amplifiers," *IEEE Transactions on Audio and Electroacoustics, AU-18,* No. 3 (Sept. 1970), p. 234.

12. W. MARSHALL LEACH, "Transient IM Distortion in Power Amplifiers," pages 34-41 *Audio* (Feb. 1975).

13. EERO LEINONEN, MATLI OTALA, and JOHN CURL, "A Method for Measuring Transient Intermodulation Distortion (TIM)," *JAES, 25,* No. 4 (April 1977), p. 170-177.

14. DON and CAROLYN DAVIS, *Sound System Engineering.* Howard W. Sams & Co., Inc., Bobbs-Merrill, Indianapolis, 1975.

10

RECORDERS
AND
RECORDING SYSTEMS

10.1 CHART RECORDERS—CONCEPTUAL REMARKS

It has been over 75 years since production and research investigators have required some record of mechanical, chemical, electrical, and/or physiological events. Today, many technological industries require a permanent record of time-varying signals. This requires electronic measurements to be made. A simple method of viewing such a signal (i.e., a plot of transistor characteristics) is to use an oscilloscope and view the scope waveform or record it with a camera. An electronic instrument which will perform this measurement and make available a permanent record of the waveform is called a graphic recorder. The graphic recorders are often called oscillographs, oscillographic recorders, strip chart recorders, or simply recorders. We shall refer to these recorders as oscillographic recorders. Oscillographic recorders find wide application in much scientific work. The need for recording systems finds its way into medicine, industrial measurements, nuclear and geological investigations, oceanography, and aerospace telemetry. These are only a few of the applications. As an example, the electrical activity of the heart can be picked up by an electrocardiograph, which is basically a recorder. The brain waves can be reproduced by a multichannel chart recorder known as an electroencephalograph.

While the oscillographic recorder has an advantage over an oscilloscope in that a permanent display of a signal is available, its main disadvantage is that the recorder can only measure signals which vary at a slow rate, with a maximum signal frequency being on the order of several hundred hertz. An oscilloscope can measure signals on the order of several hundred megahertz.

A strip chart recorder works well for direct current but is insufficient for measuring fast responses or alternating-current signals. The oscillographic recorder works well up to approximately 200 Hz.

In industrial applications, the oscillographic recorder serves to aid in the design process, to help in troubleshooting, and to assist in the control of a product.

The oscillographic recorder consists of the following basic components:

1. An electromechanical device to convert an electrical input signal to a mechanical movement.
2. A stylus leaving a written record on chart paper as the stylus moves across the paper.
3. A chart paper assembly consisting of a chart paper supply roll with an associated drive mechanism to move the chart paper across the writing table.

The heart of any recording system is, therefore, the pen, the pen motor, and the paper and ink which give the recording performance.

Many recorders have internal signal conditioning units (amplifiers) to enlarge the signal so that the excursion of the stylus will be large enough to provide a usable permanent written record.

The basic factors in selection and use of a recorder are

1. Frequency response.
2. Sensitivity (damping and power).
3. Range.
4. Accuracy.
5. Type of presentation.
6. Cost.
7. Accessories.
8. Weight, size.

Recorders may be classified into two basic types: galvanometric and potentiometric. See Fig. 10.1.

10.1.1 Galvanometric Recorders

Two types of recorders use the galvanometric principle: pen and light-beam types.

10.1.1.1 Pen recorders.   In galvanometric recorders, the pen or stylus is attached to a coil in the field of a permanent magnet. See Fig. 10.2. Pen

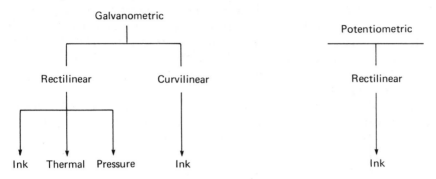

Figure 10.1 Classification of chart recorders.

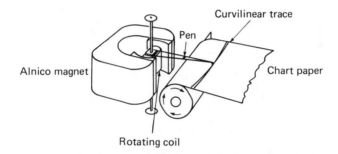

Figure 10.2 Galvanometric-type recorder.

recorders may use either (1) an ink pen, (2) a stylus tracing on carbon-coated film, or (3) a heated stylus on heat-sensitive paper. Many ink-type recorders use curvilinear coordinates on a chart because the arc of the pen is a curve, not a straight line.

Figure 10.3 shows the difference between a curvilinear and a rectilinear recording of a signal. For the recorder to measure minute electrical signals, a dc differential amplifier is usually an integral part of the recorder. Pen-type recorders may be multichannel, with a maximum frequency of 200 Hz.

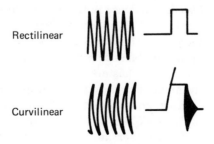

Figure 10.3 Difference between a rectilinear and a curvilinear trace.

10.1.1.2 Light-beam galvanometric recorder. The principle of the light-beam galvanometer is shown in Fig. 10.4. Since the galvanometer moves only the tiny mass of a mirror, the frequency response is increased from the 200-Hz upper response figure of pen recording types to about 5,000 Hz, which is near the practical limit of light-beam recording types. The beam "writes" on photosensitive film or paper.

In light-beam recording, both rapid wet and dry processing techniques (without chemicals) now give almost immediate readout, eliminating the once inherent disadvantage of this technique—developing time.

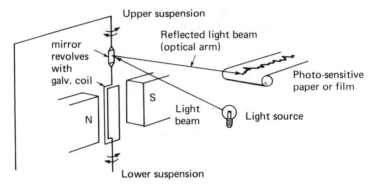

Figure 10.4 Principle of galvanometer light-beam recording.

10.1.2 Potentiometric Recorders

The potentiometric recorder operates on a servo principle, with the position of the stylus arm being deflected via a mechanism attached to the arm and in contact with a slide wire potentiometer. Due to the inherent speed of the servo system and the finite mass and frictional forces associated with the mechanism, potentiometric recorders are low-frequency devices with a frequency response limit of 10 Hz. This type may also be referred to as a null balance recorder. See Fig. 10.5 for a conventional servo potentiometric recorder.

The advantages of the null balance potentiometer are (1) high sensitivity, down to microvolt signals, and (2) independence of lead length. The sensitivity is due to the inherent amplification in the servo system. Independence of lead length is due to the null balance (servo) operation; no signal flows at balance, and thus lead resistance has no effect.

These two basic advantages are gained at the expense of response speed; at the present state of the art, potentiometers cannot operate at speeds faster than about 0.1–0.2 s of full-scale pen travel, limiting the response to signals of less than a few hertz. However, at low frequencies the potentiometer recorder is the most accurate available.

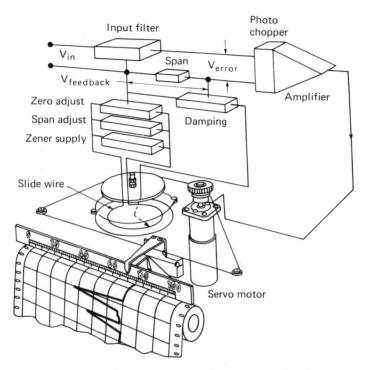

Figure 10.5 Conventional servo-potentiometer recorder uses servo motor to control the pen connected to the slide wire. Courtesy of Texas Instruments servo/riter®.

Nearly all process and test variables with a suitable electromechanical transducer can be measured in terms of microvolt or millivolt electric signals and often must be measured at great distance from the point of recording. The potentiometer recorder meets these needs. Further, the potentiometer can be used with minor modifications as a null-balancing bridge-input recorder, which is basically a resistance (or inductance or capacitance) measuring device. Hence this instrument is a general-purpose, sensitive instrument for measurement of EMF, resistance, capacitance, and inductance.

Potentiometric recorders can be used in chemical labs for liquid, electro-chemical, and thermal analysis.

10.1.3   X-Y Recorders

Conventional recorders plot one variable (the dependent variable) against time. The X-Y recorder can plot any given variable against any other variable, including time. For example, the X-Y recorder can plot the relationship between current and voltage, lift and drag, and speed and torque.

Two signals are recorded simultaneously by one pen, using the null balance or galvanometer light-beam technique. The pen is driven along the X axis by one signal and along the Y axis by the second signal. This provides rapid plotting of a dependent variable against the independent variable, such as recording tube characteristics, filter characteristics, etc. Applications of measurements will be given in Section 10.1.5.

Most X-Y recorders are self-balancing potentiometers, with either a flatbed or drum recording surface. The inputs are slowly varying dc voltages (from millivolts to volts). Capillary-type ink pens are the most common, although high-speed plotters make extensive use of point-type recordings.

Slewing speeds vary from 10 to 40 in./s, with usable writing rates being roughly one-half the slewing speed.

The X-Y recorder can combine the operations of measurement, plotting, and curve fitting within seconds. It is a slow-speed device compared to oscilloscopes and galvanometers, but it is high speed compared to manual data taking. Its advantages include

The plotting of complex curves.

Reproducible records on standard-sized graph paper suitable for direct insertion in a report.

A second independent variable may be used to draw a family of curves on a single sheet.

Parameter changes are immediately obvious by drawing families of curves on one sheet.

Self-checking: retracing the curve in the opposite direction.

The X-Y recorder is also used as an output table for analog computers. Ease of leading, accuracy, and multiple scales are features of value in this application. Whenever the relationship between two variables is desired, the X-Y recorder can supply an accurate, rapid answer. See Fig. 10.6 for an X-Y commercial recorder.

10.1.4 Magnetic Recorders

A magnetic recorder [1] is an analog storage device which stores electrical signals on magnetic tape. Although the graphic recorder also stores the electrical information, it is very difficult to regenerate the information in its original electrical format. The magnetic tape recorder is a device which readily permits recording of data in such a manner as to allow it to be produced at a later time in its original electrical format. See Fig. 10.7 for a block diagram of a magnetic recorder. The recorder also has the advantage of recording the information at one speed and then through the use of special circuitry reproducing the intact signal at a different speed. This concept enables a time expansion or compression of the information. By recording

Figure 10.6 Commercial brush 500 X-Y recorder. Uses new fiber tip pen. Has fast response and a writing speed of 40 in./s. Uses metrisite® feedback system which enforces 99.85 % linearity. Courtesy of Gould Inc., Instrument System Division, Cleveland, Ohio.

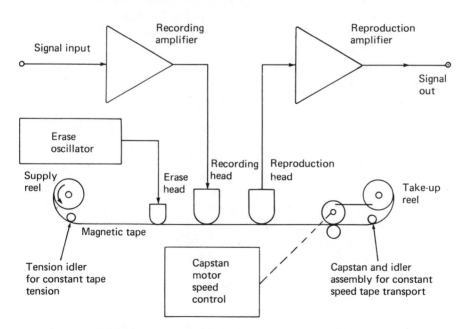

Figure 10.7 Block diagram of a magnetic recorder.

at a low speed and reproducing at a faster speed, many hours of information can be compressed and reproduced in a few minutes.

See Fig. 10.8 for a commercial tape recorder utilized in recording digital information.

10.1.5   Recorder Measurement Applications

In this section, recorders will be used to determine electronic measurements used in dc tranisient, audio, machinery data, digital, and biomedical applications. For each example block diagrams will be given when appropriate as well as a discussion of the measurements.

10.1.5.1   Basic transient dc circuit.   When the switch, S_1, is closed [position (A)] in Fig. 10.9, current will flow, charging the capacitor. The voltage across the capacitor, V_c, can be easily recorded with any oscillographic recorder which gives a rectilinear display. The type of recorder used will depend on the pen speed, the signal sensitivity, and the RC time constant (resistance times capacitance) in seconds, which determines the slope of the recorded display. An oscilloscope could have performed the same measurements with the disadvantage of not having a permanent record of the measurement.

By applying Ohm's law to the resistor in the circuit, the current through the capacitor can be calculated (current is the same through each element in a series circuit). A stop watch and a voltmeter could easily be used to determine the voltage across the resistor at any particular time and to plot current vs. time for the capacitor. It is assumed that any loading effects of the voltmeter used have been compensated for. If the switch, S1, is now placed in position (B) (across the series resistor and capacitor combination), a measurement of the voltage discharge across the capacitor is performed and is seen on the readout device. The voltage across the capacitor will drop from its reference, the battery voltage, to zero in an amount of time depending on the discharge RC time constant. The resultant voltage across the capacitor will be the sum of these two effects. The capacitor voltage will rise from zero to a maximum (battery voltage) and then will fall to zero.

10.1.5.2   Electronics.   A typical important electronic measurement to be made is the determination of the Zener diode characteristics, as shown in Fig. 10.10. Such a measurement can also be made for the conventional rectifier diode if the diode polarities and the power supply polarities are reversed. The current through the Zener diode in Fig. 10.10 is obtained by measuring the voltage drop across R_1, equal to 1 Ω. By using Ohm's law,

$$I = \frac{V}{R_1}, \qquad I = \frac{V}{1}, \qquad I = V$$

the voltage, V, is equal to the current, I, through the resistor, R_1. An X-Y recorder connected as shown in Fig. 10.10 will obtain the Zener diode

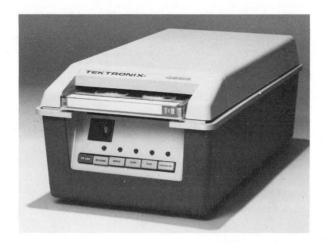

Optional Accessories

Service Manual (070-1909-00)
Replacement tapes, pkg. of 5
(119-0680-01)

Specifications

Cartridge type:
DC300A 3M® data cartridge
Tape length:
300 usable feet
Storage capacity:
200,000 bytes (nominal)
Characters/record:
128 eight-bit bytes
Recording density:
1600 bpi
Data transfer rate:
Internal—48Khz
External—Standard, up to 10k baud
Option 1, 110 to 9600
baud selectable
Data format:
8-bit binary or 8-bit ASCII
Data integrity:
Performs a read-after-read error
check when an error is detected
Number of tracks:
One effective data track

Standard Accessories

Users manual (070-1908-00)
One data cartridge

Recording format:
NRZ two-track self-clocking
Read/write speed:
30 ips
Skip forward/reverse:
30 ips
Fast forward/rewind:
90 ips
System error rate:
10' or greater
Start time read/write:
25 ms
Start time fast forward/rewind:
75 ms
Transmission characteristics:
Standard—depends on terminal
environment
Option 1—full duplex

Power requirements:

100VAC to 240VAC, 50 Hz to
60Hz, jumper selectable
Power consumption:
62 watts at 115VAC, 60Hz
Dimension:
Width—8.75 inches
Depth—17.25 inches
Height—6 inches
Weight—17 pounds

Figure 10.8 Commercial digital tape recorder. Courtesy of Tektronix Inc., Beaverton, Oregon.

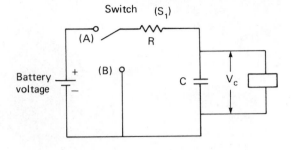

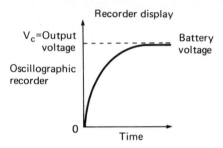

Figure 10.9 Voltage display across a capacitance with switch (S_1) in position (A). Block diagram of measurement hookup is shown on the left.

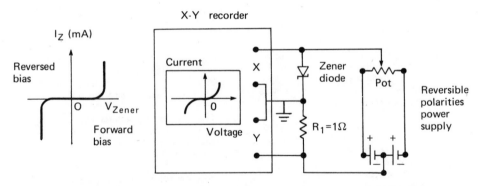

Figure 10.10 Zener diode characteristic measurement using an X-Y recorder.

characteristics. The Y axis of the recorder is set at the full-scale deflection of a few millivolts, so that a few milliamperes through R_1 will produce a readable recording.

As the potentiometer in Fig. 10.10 is rotated, the complete Zener diode characteristics can be drawn.

10.1.5.3 Audio.   The method which AM (amplitude-modulated) broadcasting stations utilize in limiting the percentage modulation to near 100%

is called audio compression. The method of evaluating an audio compressor [2] is the plotting of the program material by means of a level recorder at the input and output of the device. Although this kind of information is highly valuable, it does not allow us to make a numerical estimate of the increase in broadcast coverage area obtained through audio processing. The range is directly related to the radiated power. Accordingly, the objective of the measurements will be to exactly calculate the radiated power figure obtained over the same time period with and without processing of the signal. The ratio between these two figures will indicate the power increase obtained. The circuit arrangement used in the tests can be seen in Fig. 10.11. The test signal is a section of the normal programs used for broadcasting. This is the only method in which the evaluation can be completed, since no artificial signal would produce results similar to the real ones due to the

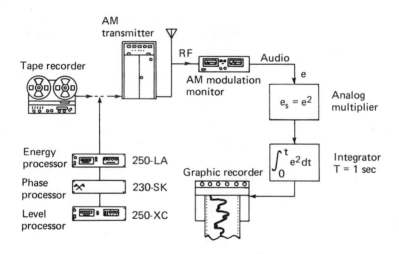

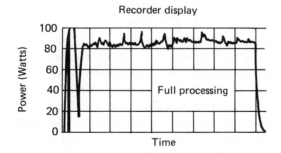

Figure 10.11 Circuit diagram for audio compressor measurements. Courtesy of *JAES*, Vol. 24, No. 5, June 1976, p. 382.

dynamic characteristics of the processors. The program material, coming from a tape recorder, is fed to a transmitter without processing or with different degrees of processing. The program, with a duration of 3 min, includes music, speech, and intervals according to current broadcast practice. At the beginning a pure audio tone is included as a reference for the adjustment of 100% modulation. The transmitter signal is applied to a modulation monitor and is then fed to an analog multiplier, which squares it, since power is proportional to the squared voltage value. After the multiplier, an electronic integrator is inserted in order to obtain average power indications on the recorder. The system gain is set so that the 100% modulation coincides with the maximum excursion of the recorder, nominally calibrated as follows: power = 100 W. The real scale has no importance since the magnitude to be measured is the power ratio.

10.1.5.4 Machinery data.   Machinery data can be analyzed on magnetic tape.

Figure 10.12 shows how to analyze data previously recorded on magnetic tape. Speeding up the tape recorder also serves to reduce total analysis time. The fixed bandwidth analyzer analyzes the frequency response between specified limits.

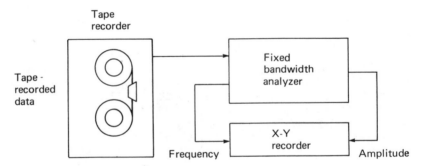

Figure 10.12 Analysis of taped data (tape recorded at 30 in./s) and speeding up the recorded data to 60 in./s. Courtesy of Spectral Dynamics Corp. of San Diego, San Diego, Calif.

A technique for automatic speed normalization of a spectrum display for use with tracking-type filter analyzers is defined as the signature ratio. The signature ratio [3] concept uses the technique of varying the sample rate to the real-time analyzer in accordance with some external influence such as a machine running speed determined by a tachometer signal. Figure 10.13 shows the measurement connections for using a signal ratio adapter with a narrow-band real-time analyzer (spectrum analyzers). The adapter varies the analysis sampling rate and the analysis band-pass filter bandwidth as a function of the tachometer input. This controls the analysis range of the

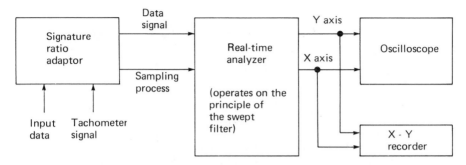

Figure 10.13   Signal ratio analysis measurement. Courtesy of Spectral Dynamics Corp. of San Diego, San Diego, Calif.

real-time analyzer. The resultant display consists of a spectrum display over a preselected number of harmonic orders.

10.1.5.5   Microwaves.   Microwave frequencies are classified as frequencies above approximately 1 GHz (1,000,000,000 Hz). To obtain efficient transfer of microwave power to an antenna, receiver input, or other load, one must match the load impedance to that of the energy source or the transmission line. Voltage or current discontinuities caused by reflection of energy are characteristic of an unmatched transmission line and can be measured by determining the standing wave ratio (SWR). Refer to section 13.6 for a more detailed discussion of SWR.

Because source SWR [4], like directivity, causes uncertainty in high-frequency measurements, it is a useful term to quantify. A system for measuring the output SWR of microwave sweepers utilizing an X-Y recorder is shown in Fig. 10.14(a). The recorder display is shown in Fig. 10.14(b).

10.1.5.6   Digital systems.   Signals from a digital machine may be plotted and shown on an X-Y recorder. Figure 10.15 shows the main digital system elements being fed into an X-Y recorder. In each instance, the digital information is fed to interfacing equipment to be sent to a digital to analog converter for analog readout in terms of voltage on the X-Y recorder. The computer sources could be keyboard, paper tape, punched cards, magnetic tape, teletype receiving terminals, computer readout, and calculator readout.

Recorders can be coupled to digital systems by the following techniques:

1.   The recorder may be an analog recorder, responding to voltage levels on each axis of an X-Y recorder. Voltages for actuation of X and Y motions are derived from digital to analog converters as shown in Fig. 10.15.

2.   The recorder is designed particularly for digital systems and is fitted with step-by-step drive systems; data enter a digital recorder at a bit rate suitable for the stepping speed of the digital drive systems.

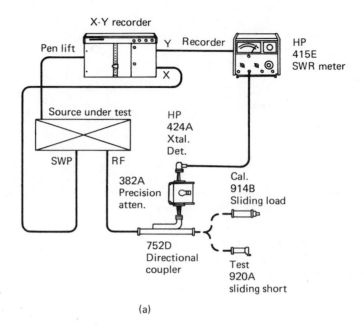

(a)

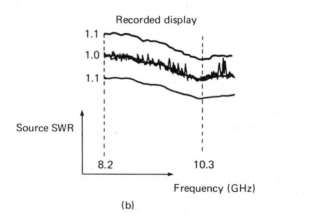

(b)

Figure 10.14 (a) System for measuring the source SWR of sweepers with final output in the waveguide. (b) Recorded display. Courtesy of Hewlett-Packard, Palo Alto, Calif.

Signals from a computer system can also be converted to an analog signal on an oscilloscope or on an oscillographic recorder.

10.1.5.7 Electrocardiogram.   The mechanical pumping action of the heart is synchronized by electrical signals which are emitted from certain nerve fibers. By monitoring these electrical signals a doctor may receive a prewarn-

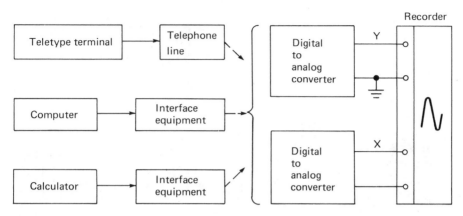

Figure 10.15   Digital recording using an X-Y recorder.

ing of any heart malfunctioning. The amplifying and recording of these signals is referred to as an electrocardiogram (ECG). The instrument utilized in monitoring and recording these signals consists of several electrodes, a high gain differential amplifier, and an oscillograph for a permanent recording of the signals. To reject any 60-Hz noise, the differential amplifier should have a large common mode rejection ratio. Figure 10.16 shows a block diagram of a typical electrocardiograph.

10.1.5.8   Electroencephalography.   Electroencephalography (EEG) is the study of the electrical activity of the brain. This electrical activity is recorded from electrodes placed on the scalp or by implanting electrodes within the brain. The low-frequency output signals (0.5–100 Hz), which are on the order of 50 μV, are amplified and displayed on a graphic recorder, with the recorded output called an electroencephalogram. The EEG instrument usually consists of as many as 6 channels, with a control switch to monitor as many as 16 different electrodes placed around the scalp and head. An analysis of these signals is helpful to a neurologist in detecting any brain malfunctioning. See Fig. 10.17 for a typical EEG recorded display.

In recording EEG signals, more care must be taken than when recording ECG signals because EEG signals are smaller (2–100 μV compared to 50–5,000 μV for the ECG), more complex, and nonperiodic and contain frequencies with continually shifting phase and amplitudes. Signals which are buried in the noise can be retrieved by digital signal averaging. As the EEG signal is not repetitive, recordings must be taken for a long period of time to obtain useful data.

Signal conditioners (amplifiers) and recording equipment required for EEG are basically the same as for ECGs. The frequency content of the two signals is approximately the same (0.05–100 Hz compared to the ECG

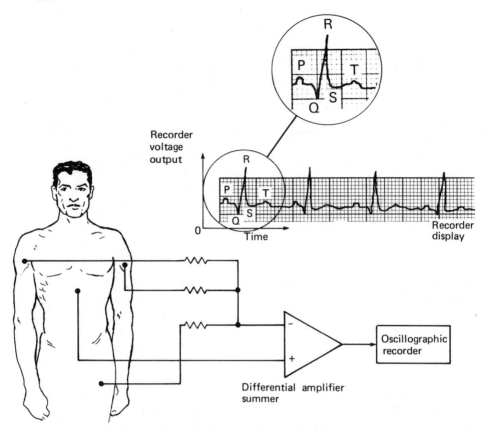

Figure 10.16 Typical electrocardiogram.

frequency content of 1–100 Hz), but, due to the smaller voltage of the EEG signal, more gain is required (approximately 1,000–10,000).

The same recorder type can be used, with usual paper speeds of 5, 15, 25, or 30 mm/s. Further use of recorders in biomedical instrumentation is described in Section 10.2.

10.1.5.9 Railroad track inspection.   Each spring and fall an instrumented rail inspection car developed by the Chessie System is pulled at posted speeds over the company's entire 11,500 mi of right-of-way. Inside the car, a Brush 200 direct writing recorder with incremental drive provides a permanent analog chart record of important track parameters on every foot of track [5].

A pulse generator driven from an axle increments the chart drive in synchronism with car movement on the scale of 1 in. of chart paper per 400 ft of track. Observers in the car actuate event markers to identify mileposts and other features on the chart for later use in directing maintenance crews to trouble spots in the track or roadbed.

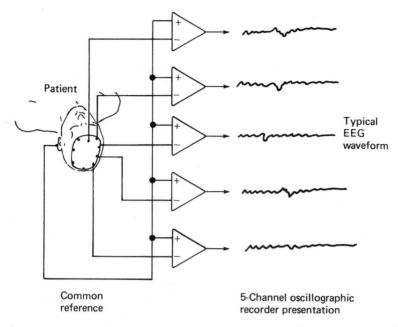

Patient

Typical
EEG
waveform

Common
reference

5-Channel oscillographic
recorder presentation

Figure 10.17   Typical electroencephalogram.

Sensors underneath the car continuously feed analog data to the recorder. Track parameters that are recorded at car speeds up to 80 mi/h are track curvature, cross-level or roll axis, vertical displacement of each rail (including joints), and track gage.

So accurate are the chart records produced by the Brush 200 that it is common for track crews to find specified faults within a few feet of where engineers directed them by studying the chart representations.

The Chessie System uses the permanent chart records to (1) detect dangerous track conditions requiring immediate attention, (2) allocate maintenance funds by territories, (3) schedule track maintenance work, (4) verify the quality of maintenance work done, and (5) correlate track conditions with traffic loads and speeds.

This application also is an excellent example of the use of an incremental drive to advance the chart paper of a recorder according to events (pulses representing distance in this case) rather than at a constant speed determined by an internal chart drive.

10.1.5.10   Monitoring noise in a moving vehicle.   Permissible noise exposures for occupational workers are now clearly defined and limited by federal law under the Occupational Safety and Health Act (OSHA), administered and enforced by the Department of Labor. Under this act, unprotected sound levels for various exposure times presently may not exceed the following:

Exposure Limit h/day	Sound Level dB (A)
8	90
4	95
2	100
1	105
½	110

A mobile instrumentation system [6] used by a manufacturer of materials handling vehicles measures and records sound levels in decibel-amperes as a function of time at the operator's ear level. The system is applicable to noise monitoring of all kinds of mobile equipment. The permanent chart records that result are useful for OSHA compliance and noise abatement studies.

Noise tests are made as the gasoline-powered vehicle traverses a standard test course carrying its full-rated load and performing a set routine of materials handling operations. A condenser microphone with a wind screen mounted at the operator's ear level picks up sound. Associated equipment is mounted on a carrier attached to the side of the vehicle. Instrument power is supplied by a dc to ac converter operated from the vehicle's storage battery. Microphones are also discussed in Chapter 11.

Output of the microphone is fed to a sound level meter which provides an amplified sound pressure level signal. This signal goes to a log converter that provides an analog sound power signal in dB (A) relative to 10–12 W. The dB(A) signal is recorded on one channel of a general-purpose (Brush 260) recorder. Other channels record engine revolutions per minute and manifold vacuum on the same time base for correlation of noise with engine speed and load.

The sound level trace is analyzed, and the data are used to calculate an overall exposure ratio reflecting the actual versus the permitted exposure times for noise levels of 90 dB(A) and above. If the exposure ratio is less than unity, the equipment can be used continuously for an 8-h shift without operator protection.

See Fig. 10.18 for a block diagram of instrumentation carried by a vehicle for monitoring noise levels dB(A) and the sound level recording made on a Brush 220 recorder.

There are hundreds of applications where a permanent record of a measurement is required. We have touched on only some of the important measurements in various electronic areas.

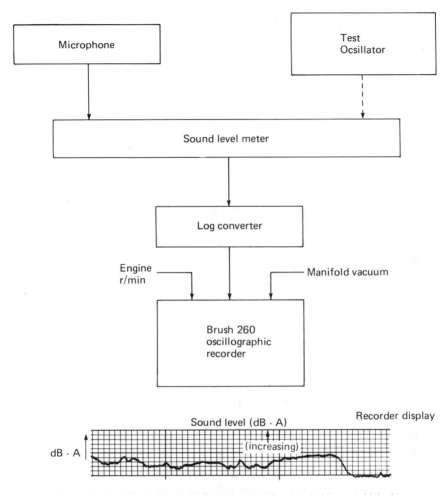

Figure 10.18   Block diagram of instrumentation carried by a vehicle for monitoring noise levels in dB(A). Courtesy of Gould Inc., Instrument Systems Division, Cleveland, Ohio.

10.2   RECORDING SYSTEMS MEASUREMENTS

In any electrical or mechanical event specifically one form of energy can be transformed to another. The signal is amplified and then *read out* with a chart recorder, oscilloscope, meter, or magnetic recorder. Today we try to analyze electronic measurements and come up with an accurate answer. For this reason, we have chosen to deal with bioelectronic recording systems. The most complex electronic situation consists of the way the human body works. Chart pen recorders were used in cardiovascular sound investigations as early as 1893. There is no doubt that chart recorders, as crude as they were, were used before that date.

10.2.1   Bioelectronic Recording System Concepts*

Physiological studies in experimental animals and man require the continuous and simultaneous recording of a variety of dynamically changing variables. Modern technology permits the transduction of almost any physiological variable into an electrical signal which can be amplified and used to drive a suitable recorder. Advantages of such electrical recording techniques include high sensitivity and excellent dynamic response which permit faithful measurement of "the quick and the small." In addition, these modern recording systems possess great versatility and convenience. Thus, given a suitable transducer (that is, a device for changing one kind of energy into an electrical voltage), the same recorder can be used to measure pressure, flow, temperature, oxygen saturation, etc. It can also be used without a transducer to measure those physiological variables which are electrical potentials or *biopotentials.*

See Fig. 10.19 for various biopotential signals measured in man, and see Fig. 10.20 for a block diagram of a physiological recording system.

We shall briefly examine each of these blocks in turn:

1.   The *physiological input signal* may be pressure, flow, temperature, a biopotential, or some other physical quantity existing at a particular point in the organism.

2.   A *transmission line* is required for certain parameters only if the input transducer cannot be or is not placed directly at the desired point of measurement. For example, if we wish to measure pressure in the right ventricle of the heart, we may not introduce our transducer directly into the ventricle. Instead a tube (catheter) is inserted into the ventricle, and it connects to the transducer, which is located some distance away. Such a transmission line is frequently the weakest link in the entire system, and the overall speed of response of the system can be no better than its weakest link. Thus, if the rest of the system can faithfully follow a change occurring in 0.01 s, it would be improper to use a compliant transmission line which could not follow a change occurring in 0.5 s. Hence, some knowledge of the characteristics of such lines and the characteristics of the phenomenon being measured is essential if we are not to fool ourselves and if we are to fully exploit the capabilities of the rest of the system. Another example of a transmission line is the mechanical or electronic stethoscope.

3.   The *input transducer* changes some form of energy such as mechanical or heat into electrical energy so that an electronic voltage amplifier may amplify the input signal. The nature of the input transducer will depend on

*Section 10.2.1 is based on application notes courtesy of Beckman Instruments Inc., Schiller Park, Ill.

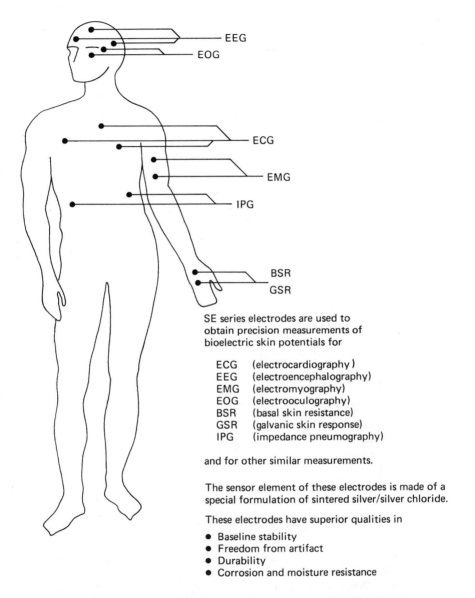

SE series electrodes are used to
obtain precision measurements of
bioelectric skin potentials for

ECG	(electrocardiography)
EEG	(electroencephalography)
EMG	(electromyography)
EOG	(electrooculography)
BSR	(basal skin resistance)
GSR	(galvanic skin response)
IPG	(impedance pneumography)

and for other similar measurements.

The sensor element of these electrodes is made of a
special formulation of sintered silver/silver chloride.

These electrodes have superior qualities in

- Baseline stability
- Freedom from artifact
- Durability
- Corrosion and moisture resistance

Figure 10.19   Biopotential skin electrodes. Courtesy of In Vivo Metric
Systems, Redwood Valley, Calif.

the nature of the physiological variable being measured: Refer to Chapter 11
for principles of transducers.

    a.   If it is a biopotential, no transducer is required; only suitable
pickup electrodes are needed.

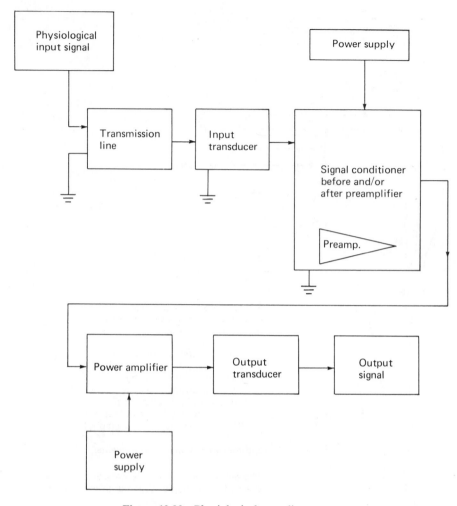

Figure 10.20   Physiological recording system.

  b. If it is a mechanical quantity (pressure, flow, etc.), a suitable electromechanical transducer is required. These may be strain or pressure gages (a Wheatstone bridge), a differential transformer, or various other devices.

  c. If it is temperature, a suitable thermoelectrical transducer is needed. This may be a thermocouple (two dissimilar metals in contact which have a potential difference at the contact junction that changes with temperature). The potential difference is on the order of microvolts. A thermistor is a variable resistor whose resistance changes with temperature. It is put in a Wheatstone bridge circuit whose output is much greater than the signal from a thermocouple. See Chapter 11 for further details.

    d. If it is the optical density of blood (as in oximetry, using a Waters ear oximeter which has a photocell and light source to measure the light absorbed by the blood passing through the ear between the photocell and the light source, which changes with the oxygen content of the blood; or as in dye dilution studies using our cardiodensitometer for cardiac output) a suitable optic-electric transducer is required. This is usually a photo-electric cell.

Whatever the physiological input to the transducer may be, the output is an electrical signal (voltage) which is quantitatively related to the input.

    4. The essential function of the *signal conditioner* is to extract from the incoming signal information to form a useful signal waveform. It may filter the incoming signal to obtain information in a desired frequency segment, or it may also yield a signal whose amplitude is proportional to some parameter of the incoming signal, for example, heart rate from the ECG or blood volume from blood flow. In addition to its primary function of signal conditioning, the signal conditioner often performs a number of accessory functions. Thus it might supply excitation voltage for strain gage bridges or thermistors. Calibration and balance controls are generally an integral part of the signal conditioner circuitry. The signal conditioner may be a separate module, or it may be an integral part of the amplifier.

    5. The function of the *preamplifier* is to increase the magnitude of the signal voltage from the signal conditioner when needed; that is, it multiplies the input voltage by a constant, K equal to or greater than 1, where K is the gain of the amplifier. In so doing, it draws negligible power from the input source. When recording the output from other devices where the signal detection, conditioning, and amplifying are done in the other device, the overall gain of the preamplifier can be set to less than 1 to accept signals that are "too high." This can be done by placing a voltage divider at the input of the amplifier.

    6. The essential function of the *power amplifier* (driver amplifier) is to provide power amplification to levels sufficient to drive the output transducer. Although transistors (or vacuum tubes) are used for both voltage and power amplification, the circuitry and the transistor or tube characteristics are somewhat different depending on which is desired. A power amplifier generally requires a relatively high input voltage signal—and hence the necessity for preamplification.

    7. The *power supply* uses a 110-ac 60-Hz source from the wall socket to supply bias voltages in the proper form to the amplifiers which drive the output transducer.

    8. The *output transducer* (recorder, readout device, data display device, etc.) may be one of several varieties, and the choice depends primarily on the rate of change of the physiological variable being studied. The ability of a measuring instrument to follow dynamically changing signals is most often expressed in terms of *frequency response*, that is, the fidelity with which sinusoidal input signals of various frequencies can be reproduced. Thus an output transducer which has a "frequency response flat from 0 to 100 Hz or c/s" will

exactly reproduce the relative amplitudes of all input frequencies within this range. Hence it is convenient to divide output transducers into four major categories depending on their frequency response, that is, on their ability to follow rapidly changing input signals:

a.   Low-frequency recorders (0–2 Hz): These recorders may consist of a direct writing coil galvanometer or a servo system.

b.   Intermediate-frequency recorders (0–200 Hz): These recorders also consist of direct writing moving coil galvanometers. However, to increase the frequency response, much stiffer restoring springs are used, and this in turn increases the power required to drive the unit (perhaps 8–10 W). Such units are often called *pen motors*. Except for very rapid phenomena such as the electrical activity of muscle or muscles action (EMG), heart sounds, or nerve action potentials, these recorders are quite generally useful for physiological studies.

c.   High-frequency recorders (0–10,000 Hz): These recorders employ optical galvanometers consisting of a very tiny mirror mounted on a very light moving coil. A recently developed instrument (Honeywell Visicorder) uses an ultraviolet light and special paper which does not require processing and makes the records available within a few seconds.

d.   Ultrahigh-frequency recorders (0–100 MHz or more): This is the familiar cathode ray oscilloscope discussed in Chapter 5.

10.2.2   Bioelectronic Instruments Using Recorders

The energies generated by the human system are as varied as the instruments used to measure them. Table 10.1 shows particular instruments and their related physiological parameters.

The signal range and frequency range given in Table 10.1 are average values. The electrical signals obtained from man or animal vary from a few microvolts to a few millivolts. Small variations in body potentials and impedance must be measured, displayed, and recorded accurately for meaningful results.

10.2.3   Biomedical Instrumentation Measurements

The most important biomedical instruments [7] used today are the electrocardiogram (ECG) and electroencephalogram (EEG). They have been described in Sections 10.1.5.7 and 10.1.5.8. More sophisticated systems are now discussed.

A highly instrumented trauma research unit in the Clinical Studies Center of Albany Medical College, Albany, New York, specializes in the treatment and monitoring of seriously injured patients, primarily accident victims.

Heart of the trauma unit monitoring instrumentation includes an 8-channel monitoring oscilloscope, an 8-channel dry-writing oscillograph, and 16 channels of biophysical signal conditioners. Two full-time biomedical engineers are on 24-h call to operate the system during care of a patient.

Table 10.1

Some Common Physiological Signals

Physiological Signal	Typical Signal Range	Frequency Range
Heart potential (ECG)	50 μV to 5mV	0.50–100 Hz
Brain potential (EEG)	2–10 mV (scalp)	1–100 Hz (scalp)
Muscle potential (EMG)	20 μV to 10 mV	10 Hz to 2 kHz (needle electrode) 10 Hz to 10 kHz (gross electrode)
Electrooculogram (EOG)	10 μV to 4 mV	0.1–100 Hz
Blood pressure pulse	5–15 m/s	
Blood pressure (indirect measurement)	0–300 mmHg	0.1–500 Hz
Blood pressure (direct measurement)	0–40 mmHg (venous) 0–300 mmHg (arterial)	
Blood flow	1–300 cc/s	
Heart sounds (PCG)		5 Hz to 4 kHz
Respiration rate	500 cc/air; 10–20 times/m	
Breath flow rate	3–200 l/m	
Untreated skin resistance	50–800 kΩ	

Patients susceptible to traumatic respiratory stress are those with an internal injury, a minimum blood loss of 1 liter, and/or a long bone fracture.

See Fig. 10.21 for a picture of an engineer, a patient, and monitoring instrumentation.

Among the measurements that can be permanently recorded on the eight-channel Gould 480 direct writing recorder are arterial blood pressure, pulmonary artery pressure, pulmonary capillary wedge pressure, central venous pressure, airway pressure, respiratory gas concentrations, temperature, tidal volume, tidal flow, ECG, EEG, arterial dye concentration

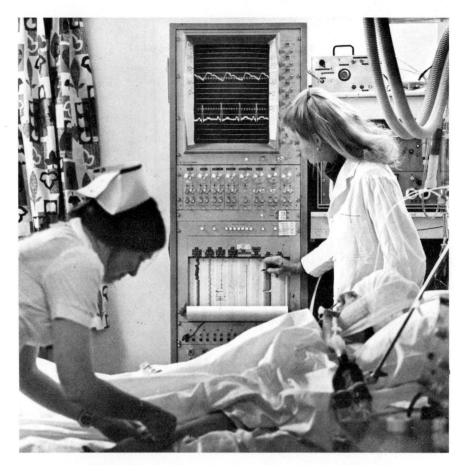

Figure 10.21 Ms. Chandler Ralph, biomedical engineer with the Albany
Medical Center trauma unit, watches for changes in the chart record
produced by an 8-channel recorder as a monitored patient receives treatment. Courtesy of Gould Inc., Instrument Systems Division, Cleveland,
Ohio.

(for cardiac output determination), and impedance of a limb segment (for
limb blood flow monitoring).

See Fig. 10.22 for a recording using the Gould 480 recorder on eight
channels of parameters on a common time base.

In Fig. 10.23, a complex block diagram of a data collection and data
processing system of kinematic measurements in walking is presented. The
data are recorded via transducers and fed to a signal conditioning device.
A tape recorder is used for data playback and processing, and this information can be sent to a computer with a display oscilloscope. Note that the
recording and reproducing systems are given.

In considering any recording measurement system, ground loops [8] are

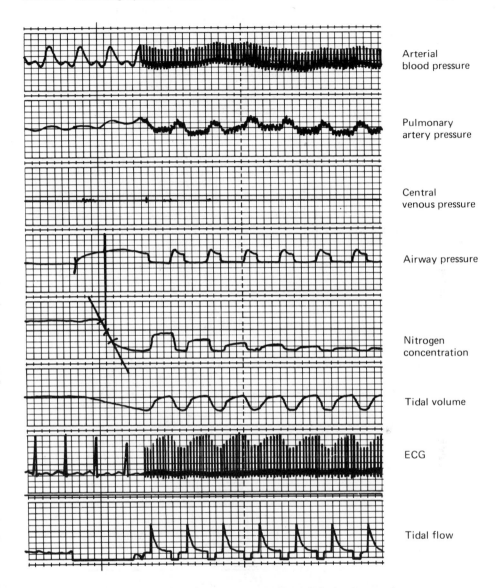

Figure 10.22 Typical patient parameters recorded simultaneously on the 8-channel Gould 480 recorder in a trauma unit. Courtesy of Gould Inc., Instrument Systems Division, Cleveland, Ohio.

the largest source of electrical noise between electronic modules. More than one ground on the signal circuit or signal shield produces a common impedance or ground loop between these two points.

Guidelines on grounding include the following important rules:

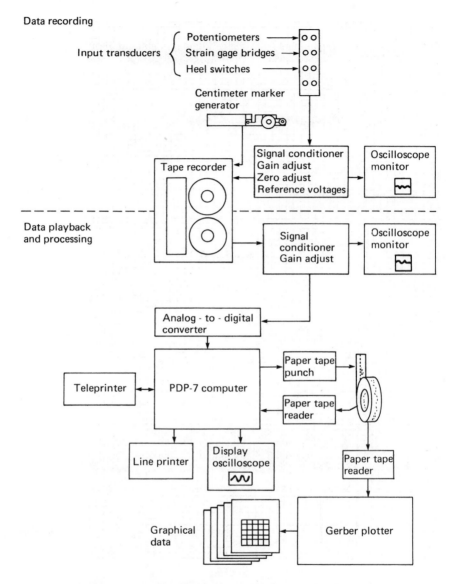

Figure 10.23 Block diagram of a data collection and data processing system. Courtesy of *Bulletin of Prosthetic Research, BPR 10-15*, Spring 1971, p. 21.

1. The recording system should have a stable system ground.
2. The signal circuit should never be grounded at more than one point.
3. The signal cable shield should not be attached to more than one ground, and this ground should be at the signal source.

4. More than one intentional or accidental group on the signal circuit or signal cable shield will produce excessive electrical noise in any low-level circuit.

5. In off-ground recording, the signal cable shield should *not* be grounded; it should be connected to the center tap or the low side of the signal source.

10.3 SUMMARY

Most laboratory and production recorders are electrically actuated. They convert an electrical signal into a displacement of the writing pen. They have three general classifications: galvanometric, oscillographic, and potentiometric. The measurement depends on the pen speed and the electronic signal amplitude.

In Section 10.2 emphasis was placed on the total transduction system, which consists of the sensor (transducer) signal conditioner and the readout device.

Variables such as temperature, volt-amperes, pounds per square inch, revolutions per minute, and degrees have typical signal-type matching, shown in Table 10.2.

Table 10.2

Matching Signal Types to Recorder

Typical Variable	Typical Transducer	Signal Type	Recorder	Remarks
Temperature	Thermocouple	dc millivolts	Single channel or servo	Nonlinear scaling
Volt-amperes	—	ac volts or current	dc electrodynam-ometer	Power freq. Waveforms
Pounds per square inch	Potentiometer	Pressure	Sincle channel or servo	Pressure snubber; may be required to prevent oscillation
Revolutions per minute	dc tachometer or generator	Speed	Single channel or servo	Requires freq. to to voltage converter
Degrees	Potentiometer or synchro	Position	Single channel or servo	Applied to shaft position

Source: From "Interpretation of Chart Recordings" by J. R. Judkins, *Instruments and Control Systems,* Published by Chilton Publishing Co., Radnor, Pa., Jan. 1976, p. 53.

The important points to be considered are the interpretation of the measurement, the data, the measuring system, and the nature of the input signal, which constitute the core of a recording process.

REVIEW QUESTIONS

1. What is a chart recorder?
2. Give three examples of a chart recorder.
3. What are the advantage and the disadvantage of a chart recorder compared to an oscilloscope?
4. What is a galvanometric recorder? Describe its operation.
5. What is a potentiometric recorder? Describe its operation.
6. What is an X-Y recorder? Give an application of an X-Y recorder in industry.
7. What is a light-beam recorder?
8. What is a magnetic recorder?
9. What is the significance of a recording system in electronic instrumentation?
10. What does biopotential mean?
11. What is a transmission line?
12. What is a signal conditioner?
13. What is a preamplifier?
14. What is a power supply?
15. What is an output transducer?
16. How does an electrocardiogram work?
17. How does an electroencephalogram work?
18. How do you interpret a recording measurement?
19. What does transduction mean?
20. What transducer would you use to measure revolutions per minute?

REFERENCES

1. *Magnetic Tape Recording Handbook, Application No. 89.* Hewlett-Packard, Palo Alto, Calif., Oct. 1975.
2. OSCAR J. BONELLO, "New Improvements in Audio Signal Processing for AM Broadcasting," *JAES, 24*, No. 5 (June 1976), p. 382.
3. ANTON C. KELLER, "Real Time Spectrum Analysis of Machinery Dynamics," S/V *Sound and Vibration* vol 9, pp 14–48 (April 1975).

4. *High-Frequency Swept Measurements, Application Note 183*. Hewlett-Packard, Palo Alto, Calif., Nov. 1975, p. 23.
5. "The Railroad Track Inspection ... The Easy Way," *Data Display*, Vol. 2, No. 2. Gould Inc., Instrument Systems Division, Cleveland.
6. "Monitoring Noise in Moving Vehicles," *Data Display*, Vol. 2, No. 3. Gould Inc., Instrument Systems Division, Cleveland.
7. L. CROMWELL et al., *Medical Instrumentation for Health Care*. Prentice-Hall, Englewood Cliffs, N.J., 1976.
8. *Gould Brush Signal Conditioning, Application Notes, No. 101*. Gould Inc., Instrument Systems Division, Cleveland, Oct. 1970.

11
TRANSDUCER SYSTEMS

11.1 INTRODUCTION

In Chapter 10, transducers were defined as devices that convert one form of energy to another form of energy. There are many types of transducers for measuring many different physical parameters. The most common uses of transducers are for the measurement of light intensity or color, concentration of liquids, flow rate of gas or liquids, velocity and acceleration, pressure, force, temperature, mechanical rotation or displacement, sound measurements, and frequency, among others. All of these physical entities must be measured accurately and with repeatability in measurement results. Most transducers are relatively simple but unique devices. Transducers are conceptually easy to design but difficult to construct and calibrate. While transforming the desired parameter, the transducer should be insensitive to all other variable parameters. For example, taking pressure measurements in a high-vibration or shock environment over a relatively small temperature band will require a pressure transducer with a very low vibration sensitivity but whose temperature sensitivity would not be critical. Because of this, transducers are one of the most important elements in an electronic measurement system. They are the bridge between the parameter to be measured and the capability of measuring that parameter.

It is our intent in this chapter to introduce the technician and engineer to the principles of the operation of and the measurements made with transducers. The nature of the electrical signal emanating from the transducer depends on the basic principle involved in its design. The output may be analog, digital, or frequency modulated.

The transducer or sensor, as it is sometimes called, has to be physically compatible with its intended application. In the selection of a transducer there are approximately eight areas of consideration:

1. Operating range: chosen to maintain range requirements and maintain good resolution.
2. Sensitivity: chosen to allow sufficient output.
3. Frequency response and resonant frequency: flat over the needed range. Will the resonant frequency be excited?
4. Environmental compatibility: temperature range, corrosive fluids, pressures, shocks, interaction, size, and mounting restrictions.
5. Minimum sensitivities to expected stimulus other than the measurand.
6. Accuracy: repeatability and calibration errors as well as errors expected due to sensitivity to other stimuli.
7. Usage and ruggedness: ruggedness both of mechanical and electrical intensities versus size and weight. Who will be installing and using the transducer?
8. Electrical: length and type of cable required; signal to noise ratios when combined with amplifiers and frequency response limitations.

In Section 11.2, the most common type of transducers will be discussed. Emphasis will be on a descriptive story of what the transducers do, where they are used, and the basic measurement principles associated with them.

The most common transducers are those in which a force is applied to the system which converts the applied force into displacement. These transducers are generally classified as

Capacitive
Differential transformer
Inductive
Piezoelectric
Piezoresistive—the strain gage
Photoelectric
Signal converters
Potentiometric
Thermocouple

We shall discuss some of these as well as other transducers in the sections which follow.

11.2 PRINCIPLES OF TRANSDUCERS
AND MEASUREMENTS*

11.2.1 Strain Gage Transducer

Strain gage transducers are used to measure force, weight, pressure, flow, torque, and many other parameters. Common transduction strain gage elements are made of wire, foil, or semiconductors.

All solid materials experience strain, or physical deformation, when they are subjected to internal or external forces. Measuring strain is vital in experimental stress analysis (particularly in aircraft structures) and plays an important role in product development and manufacturing. Many transducers or pickups—such as those for sensing force, weight, torsion, acceleration, vibration, flow, etc.—use strain gages as sensing elements.

Unit strain is a useful engineering term that permits comparison of the strain characteristics of different materials. It is also a factor in the calibration of strain gages.

Average unit strain is defined as the total deformation of a body in a given direction, divided by the original length in that direction:

$$\text{average unit strain} = \frac{\text{change in length}}{\text{original length}}$$

$$\epsilon = \frac{\Delta L}{L}$$

(11.1)

Strain gages are electromechanical transducers that exhibit a change in electrical resistance with a change in strain. The sensitivity of a strain gage is expressed in terms of its gage factor (G_f):

$$G_f = \frac{\Delta R/R}{\Delta L/L}$$

(11.2)

Gage factors with semiconductors can range from 40 to 200 compared to approximately 1 to 5 for foil or metal types.

Strain gage technology [1] is highly developed. Strain gages are available in a variety of configurations and are compounded for accurately measuring the static and dynamic strain in a wide variety of materials.

In measuring systems, strain gages typically form one or more arms of a Wheatstone bridge. If ambient temperature might change, it is important to use temperature-compensated strain gages, bridge circuits, and lead wire systems. Other circuit considerations include avoiding variable contact resistance, thermocouple effects, and strain on lead wires.

Conditioning of strain gage signals for recording can be accomplished by

*Portions of Section 11.2 are largely based on "Survey of Transducers and Signal Converters," *Electronic Instrument Digest*, May 1969, pp. 31–40, published by Milton S. Kiver Publications, Inc., Chicago.

either ac carrier or dc techniques. The ac carrier method is superior in terms of stability, freedom from electrical noise, and sensitivity. The direct dc technique is superior in terms of frequency response, operating simplicity, and amplifier versatility.

Figure 11.1 illustrates the combination of a Brush universal carrier amplifier and a direct-writing recorder in a strain recording system. The waveforms show the amplitude modulation of a high-frequency carrier wave by the strain signal and the subsequent amplification and rectification of the modulated carrier signal to obtain a faithful but greatly amplified strain signal for recording. This system provides a sensitive and accurate means for measuring all types of strain.

Transducers that use highly doped semiconductors have advantages in nuclear environment applications. Because of their low sensitivity to shock and acceleration, semiconductor units find application in blast test measurements, where pressure transducers are often exposed to large shocks (above 10,000G, acceleration, force). This characteristic, combined with the absence of moving parts, makes them suitable for many applications in severe environments.

11.2.2 Temperature Transducers

Temperature transducers can be divided into four main categories:

1. Resistance temperature detectors (RTD).
2. Thermocouples.
3. Thermistors.
4. Ultrasonic.

11.2.2.1 Resistance temperature detectors. Resistance temperature detectors commonly employ platinum, nickel, or resistance wire elements, whose resistance variation with temperature has a high intrinsic accuracy. They are available in many configurations and sizes and as shielded or open units for both immersion and surface applications.

11.2.2.2 Thermocouples. The thermocouple (TC) develops an EMF that is a function of the temperature difference between its hot and cold junctions. TCs made of base or noble metals are commonly used to measure temperatures from near absolute zero to about +3,200°F, while special units are available for temperatures to +5,600°F. Shielded and bare junction varieties, with a wide choice of sheath materials (such as Iconel, stainless steel, and noble metals), are standard items and can be obtained in a number of different configurations, assemblies, and mountings.

Thermocouples [2] are accurate, easily formed, and economical and provide good thermal response. They are subject to corrosion, are generally fragile, and are frequently difficult to mount. Metal or ceramic protecting

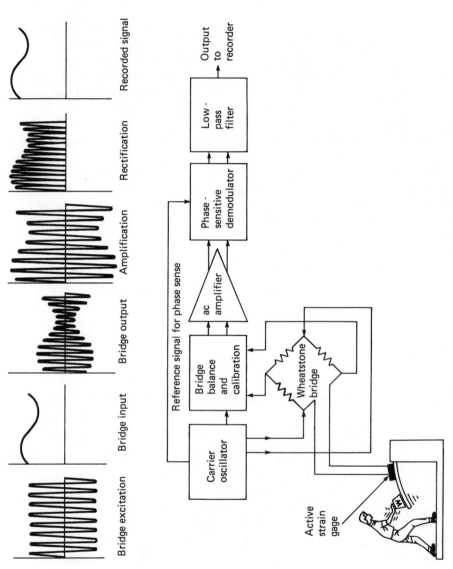

Figure 11.1 Strain gage measuring system. Courtesy of Gould Inc., Instrument Systems Division, Cleveland, Ohio.

tubes can be used to minimize these problems. To ensure accurate results, care is necessary in wiring and installing thermocouple instrumentation systems.

Most modern thermocouple systems employ electrical cold junction compensation to simulate a reference temperature. Electrical cold junction compensator cards are available for all common thermocouple materials. These cards are compact, reliable, and consistently accurate and require no operator attention or maintenance. They also are well adapted to zero suppression, an electrical technique for greatly improving resolution and accuracy in measuring small changes in temperature.

Figure 11.2 shows major elements of a thermocouple temperature measuring system with electrical cold junction compensation and zero suppression. Several readout options are indicated.

Table 11.1 gives basic thermoelectric laws and their significance in thermoelectric measurement systems.

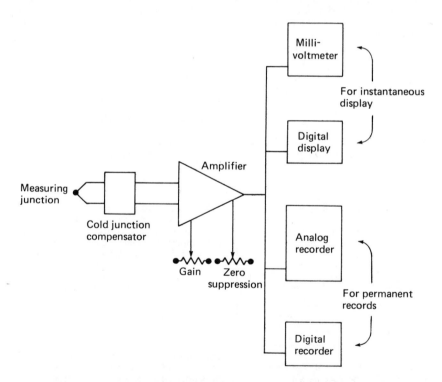

Figure 11.2 Block diagram indicates the major elements of a thermocouple temperature measuring system with electrical cold-junction compensation and provision for zero suppression. Optional readout devices are shown for both instantaneous display and permanent recording.

Table 11.1

Basic Thermoelectric Law	Practical Implications for Thermoelectric Thermometry

1. The Law of Homogeneous Circuits

An electric current cannot be sustained in a circuit of a single homogenous metal, however varying in section, by the application of heat alone.

As long as the metal of each wire in a thermocouple is homogeneous, the EMF generated at the junction will not be affected by the temperatures of the lead wires even though a temperature distribution exists along the wires.

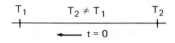

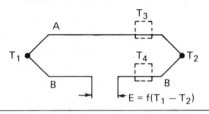

$$E = f(T_1 - T_2)$$

2. The Law of Intermediate Metals

The algebraic sum of the thermoelectromotive forces in a circuit composed of any number of dissimilar metals is zero if all the circuit is at a uniform temperature.

Any measuring device or lead wire can be added to the circuit without affecting the accuracy as long as the new junctions are at the same temperature; also permits soldering or brazing thermocouple junctions.

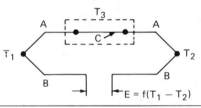

$$E = f(T_1 - T_2)$$

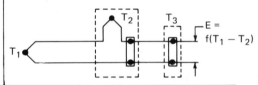

$$E = f(T_1 - T_2)$$

3. The Law of Successive or Intermediate Temperatures

If two dissimilar homogeneous metals produce a thermal EMF of E_1 when the junctions are at temperatures T_1 and T_2 and a thermal EMF of E_2 when the junctions are at temperatures T_2 and T_3, the thermal EMF generated when the junctions are at temperatures T_1 and T_3 will be $E_1 + E_2$.

This law permits the determination of measuring junction temperature from a calibration chart based on a certain reference junction temperature, when the reference junction is at a different, but known, temperature from the chart basic temperature. It also makes feasible the use of oven-controlled and electrically simulated reference junctions.

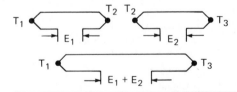

Summary

By combining these three basic thermoelectric laws, it is seen that (1) the algebraic sum of the thermoelectric EMFs generated in any given circuit containing any number of dissimilar homogeneous metals is a function only of the temperatures of the junctions and that (2) if all but one of the junctions in such a circuit are maintained at some reference temperature, the EMF generated depends only on the temperature of that junction and can be used as a measure of its temperature.

Source: Gould Inc., Instrument System Division, Cleveland, Ohio.

11.2.2.3   Thermistors.   Thermistors are semiconductors whose resistance changes with temperature. While their existence has been known for about 150 years, they did not receive extensive application until about 1940. They are commonly made of sintered oxides of manganese, nickle, copper, or cobalt and are available in disc, wafer, rod, bead, washer, and flake form, with power-handling capabilities from a few microwatts up to 25 W. Standard units have a high temperature coefficient of resistance and are produced with resistances ranging from a few ohms to 100 MΩ.

Thermistors [3] can be connected in series-parallel arrangements for applications requiring increased power-handling capability. High-resistance units find application in measurements that employ long lead wires or cables. Thermistors are chemically stable and can be used in nuclear environments. Their wide range of caracteristics also permits them to be used in limiting and regulation circuits, as time delays, for the integration of power pulses, and as memory units.

Typical thermistor configurations are shown in Fig. 11.3(a), and the electrical symbol of the device is depicted in Fig. 11.3(b).

Examples of thermistor circuits are shown in Figs. 11.4 through 11.6.

A thermistor in one leg of a Wheatstone bridge circuit will provide precise

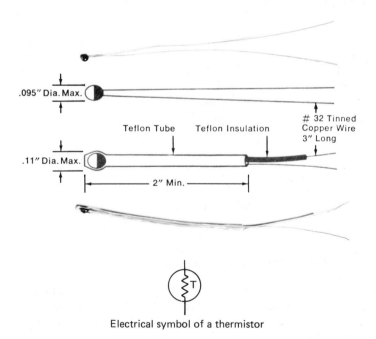

Electrical symbol of a thermistor

Figure 11.3   Thermistor configuration and the electrical symbol for a thermistor. Courtesy of Yellow Springs Instrument Co., Yellow Springs, Ohio.

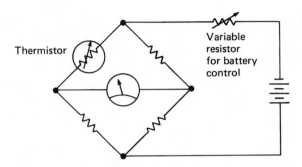

Figure 11.4 Temperature measurement circuit. Courtesy of Yellow Springs Instrument Co., Yellow Springs, Ohio.

temperature information. Accuracy is limited in most applications only by the readout device. See Fig. 11.4.

Since lead length between the thermistor and the bridge is not a limiting factor, this basic system can be expanded to measure temperature at several locations from a central point. Thermistor interchangeability and large resistance change eliminate any significant error from switches or lead length. See Fig. 11.5.

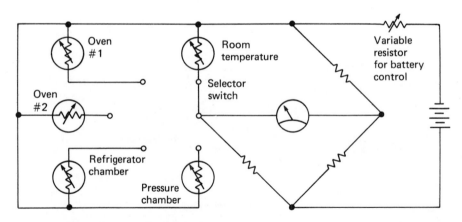

Figure 11.5 Modified temperature measurement circuit. Courtesy of Yellow Springs Instrument Co., Yellow Springs, Ohio.

Thermistors are nonlinear over a temperature range, although units today are available with a better than 0.2% linearity over a 0° to 100°C temperature range. The typical sensitivity of a thermistor is approximately 3 mV/°C at 200°C.

For accurate indication of the temperature differential, two thermistors can be used in a Wheatstone bridge circuit. Thermistor interchangeability simplifies the circuit design and reduces the number of components. See Fig. 11.6.

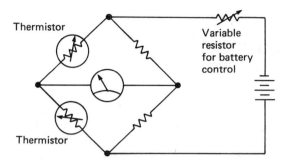

Figure 11.6 Differential thermometer. Courtesy of Yellow Springs Instrument Co., Yellow Springs, Ohio.

11.2.2.4   Ultrasonic temperature transducers.   Ultrasonics [4] (sound vibrations above 20,000 Hz) can be useful when we are concerned with rapid temperature fluctuations; temperature extremes; limited access, nuclear, or other severe environmental requirements; and measurements of temperature distribution inside solid bodies. The need to measure simultaneously the distribution of parameters other than temperature (for example, flow) may also justify an ultrasonic approach. Ultrasonics also offers possibilities of remote sensing and sometimes can avoid any penetration of the system.

Apart from profiling, ultrasonic thermometer sensors permit one to measure an extremely wide range of temperature, from cryogenic to plasma levels, and to achieve micro- to millisecond response time, millidegree resolution, greater choice of materials for sensors, operation in extreme nuclear or corrosive environments, averaging capability over a defined path, and remote location of transducer and electronics. Naturally, not all these features are available simultaneously.

Regarding profiling, ultrasonics permits one to obtain from 2 to 10 or more temperatures using a single transmission line. This feature minimizes the perturbation of the region in question, simplifies installation, and provides reliable, accurate data at a reasonable price per point.

A schematic and oscillogram illustrating ultrasonic temperature profiling is shown in Fig. 11.7. An ultrasonic thermometer fabricated by Panametrics of Waltham, Mass., is shown in Fig. 11.8.

11.2.3   Potentiometric Transducer

Linear displacement is achieved by using a pot (slide bar) or pressure using a Bourdon tube with a slide pot or even the capsule type. The motion of the slider results in a resistance change that can be made linear depending

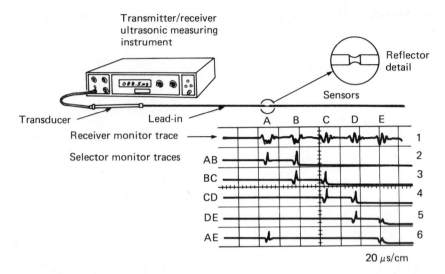

Figure 11.7 Schematic and oscillogram illustrate ultrasonic temperature profiling. Single line containing series sensors is scanned by selecting echoes according to sensor position along the line. Transit time (μs) between selected pair of echoes corresponds to temperature between reflection points. Echo pairs *AB*, *BC*, *CD*, and *DE* yield profile; pair *AE* yields average temperature. Courtesy of L. C. Lynnworth and D. R Patch, "New Sensors for Ultrasound: Measuring Temperature Profiles," *Materials Research and Standards*, *10*, (8) Aug. 1970, pp. 6–11.

on the method by which the resistance wire is wound. The trick simply is to get mechanical movement (force summing) of the work pot. The transducer limitation is resolution, physical wear with time, and shock of vibration sensitivity. A potentiometric transducer is shown in Fig. 11.9.

Rotational displacement is accomplished by utilizing a potentiometer as depicted in Fig. 11.10(a). A linear taper potentiometer with a rotation of 360° with a positive potential, *E*, at one end and ground the other end results in a reference voltage divider. If the potentiometer shaft is connected to a motor shaft, the resulting output voltage is proportional to the angle of displacement of the motor.

Refer to Fig. 11.10(b) for the potentiometer output voltage as a function of shaft rotation.

11.2.4   Variable-Inductance Transducers

The operation of variable-inductance transducers depends on the ability to measure accurately the displacement of one inductive element relative to another. Inductance transducers can be divided into two classes: those employing self-inductance and those using mutual inductance. They are

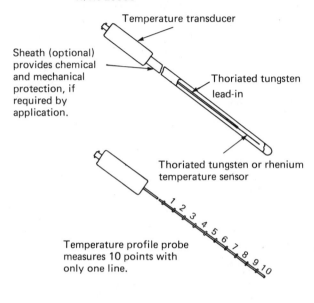

Standard high-tempreature probe
measures up to 2,500° C. Model 5010:
W/Re 2500C

Temperature transducer

Sheath (optional)
provides chemical
and mechanical
protection, if
required by
application.

Thoriated tungsten
lead-in

Thoriated tungsten or rhenium
temperature sensor

Temperature profile probe
measures 10 points with
only one line.

Model 5010-Temp Pro-10   Patented

Figure 11.8   Ultrasonic thermometer. Courtesy of Parametrics, Waltham, Mass.

further divided into those in which the inductance change is measured directly and those that provide a ratio, differential, or Wheatstone bridge output. Differential output sensors provide certain advantages, including greater output and lower susceptibility to temperature change, magnetic fields, supply voltage variations, and frequency drift.

One type of transducer, the linear variable differential transformer (LVDT), consists of three or more coils, displaced axially on a cylindrical form, with a rod-shaped core positioned to provide a path for the coil flux. When the primary is energized, voltages are induced in the series-opposing secondaries. The net output is the difference between these voltages. This output will be zero at one particular core position. Movement from this position produces a voltage that increases linearly with displacement. LVDTs are available with core displacements ranging from ± 0.005 to ± 120 in. Typical units operate with excitation voltages from 5 to 115 V rms and develop output in the range from 0.05 to 1.5 times the excitation voltage.

The LVDT transducer can, therefore, be used to measure voltage as a function of linear displacement. See Fig. 11.11(a). When the core is in the

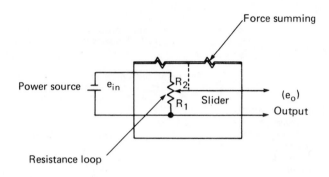

$$\text{Basic formula:}\quad e_o = \frac{R_1}{R_1 + R_2} e_{in}$$

Advantages:
- (a). high output
- (b). inexpensive
- (c). easily serviced
- (d). easy to excite and install
- (e). may be excited with ac or dc
- (f). wide range of functions
- (g). no amplification or impedance matching is necessary

Disadvantages:
- (a). usually large size
- (b). the resolution is finite in most cases
- (c). high mechanical friction
- (d). limited life
- (e). sensitive to vibration
- (f). develops high noise levels with wear
- (g). requires large force-summing member
- (h). low-frequency response
- (i). Large displacement required

Figure 11.9 Potentiometric transducer. Courtesy of Stratham Instruments Inc. From *Introduction to Transducers for Instrumentation*, p. 21, Stratham Instruments Inc., Oxnard, Calif.

neutral position, the inductance is the same on the secondary winding and the displacement will be zero and the output voltage will be zero. Moving the plunger from the neutral position to increase the amount of inductance will provide more positive voltage. See Fig. 11.11(b) for an LVDT measuring scheme of linear displacement as a function of secondary output voltage.

The Instrument System Division of Gould Inc., Instruments System Division, of Cleveland, Ohio has developed the metrisite system [5, 6], which

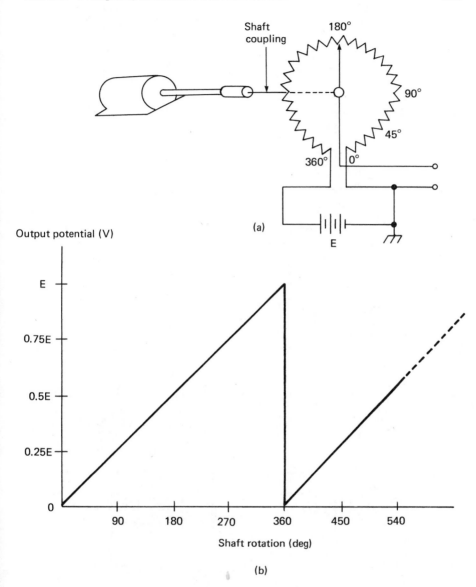

Output potential (V)

Figure 11.10   (a) Potentiometric transducer configuration. (b) Voltage as
a function of shaft rotation is shown for a potentiometric transducer.

can be used as an alternative to LVDTs and other transducers. The metrisite
is a unique type of differential transformer that translates angular position
of its input shaft into a proportional output voltage with infinite resolution.
The basic structure of the metrisite is shown in Fig. 11.12. The fixed portion
consists of a special transformer core lamination with a primary coil around

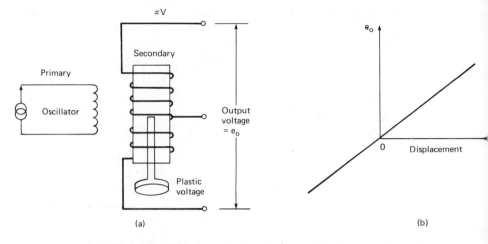

Figure 11.11 (a) LVDT transducer configuration. (b) Output voltage of the LVDT from the secondary winding as a function of linear displacement.

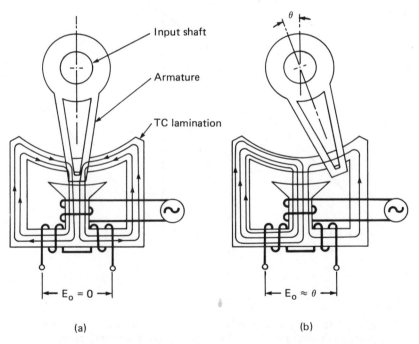

Figure 11.12 Displacement of armature from center position alters flux distribution in TC lamination to produce an output proportional to angular position of input shaft. Courtesy of Gould Inc., Instrument Systems Division, Cleveland, Ohio.

the center leg and balanced secondary coils around the outer legs. The secondary coils are connected series opposing. When an alternating voltage is applied to the primary coil, alternating magnetic flux flows through the center leg, across an air gap, and through the outer legs.

A movable armature operated by the input shaft acts as a single shorted turn in the air gap. When the armature is in its center position, the flux created by the primary excitation divides equally between the two outer legs, inducing equal and opposite ac voltages in the two secondary coils. The net output signal, therefore, is zero.

If the armature is displaced from its center position the flux distribution is altered by the circulating current in the armature. Less flux flows through the secondary coil toward which the armature is moved, and more flux flows through the opposite coil, in proportion to displacement. As a result, the secondary coil voltages are unbalanced, and net output represents the armature position.

Such a transducer is used in a servo-galvanometric recording system, shown in Fig. 11.13, which has an accuracy of 99.65%.

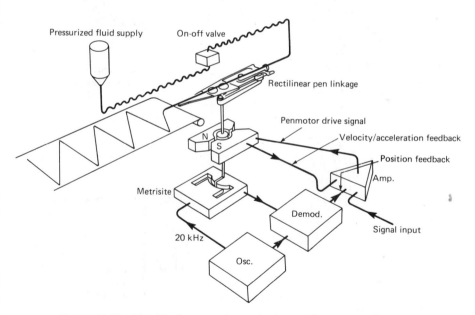

Figure 11.13   Metrisite® servo-galvanometric recording system. Courtesy of Gould Inc., Instrument Systems Division, Cleveland, Ohio.

11.2.5   *Piezoelectric Transducers*

Although the piezoelectric effect has been known for approximately 75 years, it is during the last 30 years that practical piezoelectric transducers have become common. Piezoelectricity means simply "pressure" electricity; that is, if particular types of crystals are squeezed along specified directions, an electric charge will be developed by the crystal.

See Fig. 11.14 for a bimorph construction and polarization of a ceramic element. The barium titanate wafers are soldered to the metal vane to form a sandwich and are provided with electrodes. The assembly is subjected to a strong dc polarizing voltage of about 100 V/mil applied in such a manner that the wafers are oppositely polarized. Upon removal of this dc voltage, it is found that the element exhibits piezoelectric properties. The Curie temperature is the highest temperature in which the ceramic will start having its

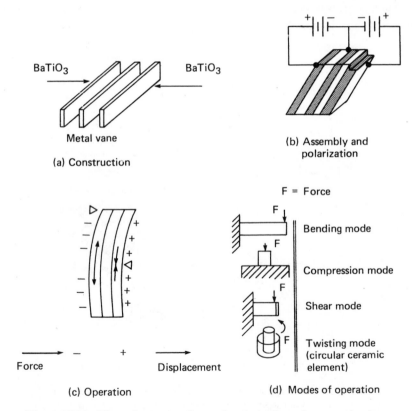

Figure 11.14   Bimorph construction and polarization of a ceramic element.

electrical properties changed, and for barium titanate it occurs at about 120°C.

Significant advances in performance characteristics of motion, force, and pressure transducers have been made in the past decade by use of new piezo materials. New transducers are available that measure high-amplitudes and also provide high-voltage outputs at small amplitudes. This is achieved by using improved sensing elements, including newer piezoelectric ceramics.

Most shock and vibration measurements are made with piezoelectric accelerometers, employed at frequencies substantially below resonance. These transducers contain a mass element and a sensor connected to the base. Acceleration motion applied to the base deflects the sensor. Since this deflection is extremely small in the case of piezo elements, the transducer is said to have no moving parts. When other sensors are used, e.g., LVDTs, the deflection is appreciable. The sensor produces an electrical output proportional to the deflection.

The charge developed in piezoelectric transducers is proportional to the piezoelectric constant of the material and to the applied stress. The constant depends on the mode of operation employed. Although quartz crystals are used in some units, the man-made ceramics are now popular, since they exhibit higher piezoelectric constants, provide higher outputs, and are less susceptible to environmental effects, such as case strains and transverse forces or motions. Lead-zirconate-titanate ceramics are used extensively, and other proprietary ceramics are employed.

Various types of accelerometers and their characteristics are listed in Table 11.2.

Piezo element accelerometers are well suited to all shock and vibration applications, including measurements requiring small or lightweight units. Units capable of measuring high accelerations have resonant frequencies as high as 250 kHz. For most applications, piezoelectric accelerometers are limited to measurements at frequencies above about 2 Hz. Rolling the response off below 2 Hz is accomplished to eliminate any pyroelectric output.

Characteristics of mechanical impedance* heads and force transducers are listed in Table 11.3.

Mechanical impedance heads are becoming more popular for analyzing the dynamic characteristics of machinery and devices with special vibration considerations. Force transducers are used to control the magnitude of vibration applied to structures being tested on shakers.

Table 11.4 presents dynamic pressure transducer characteristics.

Piezoelectric diaphragm pressure transducers are capable of performing measurements of high-pressure ranges and at high frequencies. For measure-

---

*Mechanical impedance = force/velocity. See Section 11.7 which follows.

Table 11.2

Accelerometers for Shock and Vibration Measurements

Typical Performance Characteristics

Application	Type	Sensitivity (mV/g)	Frequency (Hz)	Range (g)	Impedance
General-purpose vibration, including jet engines	Piezoelectric	18*	2–6,000	2,000	120 pF
	Piezoresistive	25	0–750	25	1,400 Ω
	Wire strain gage	0.2	0–500	100	350 Ω
	Differential transformer	100	0–50	50	500 Ω
	Variable reluctance	50	0–240	100	500 Ω
Random vibration	Piezoelectric	10*	2–6,000	1,000	750 pF
	Piezoresistive	1	0–2,000	250	500 Ω
Low-amplitude vibration	Piezoelectric	390*	2–3,000	10^{-5}–100	135 pF
	Servo	250	0–500	10^{-5}–15	250 Ω
General-purpose shock	Piezoelectric	16*	2–6,000	1,000	10 nF
	Piezoresistive	0.1	0–6,000	2,500	330 Ω
	Wire strain gage	0.04	0–2,500	500	350 Ω
	Differential transformer	1	0–500	700	—
	Variable reluctance	1	0–7,500	1,000	1,000 Ω
High-amplitude shock	Piezoelectric	0.65*	2–15,000	20,000	1 nF
	Piezoresistive	0.025	0–10,000	10,000	345 Ω

*Charge sensitivity, in pC/g. can be obtained by multiplying the voltage sensitivity by the capacitance in nanofarads.
Source: *Survey of Transducers and Signal Converters* which appeared in Electronic Instrument Digest, p. 37, May 1969. Published by Milton S. Kiver Publications, Inc., Chicago, Ill.

ment of fast rise-time pressure pulses, such as those caused by explosives, it is necessary to use an elastic wave bar transducer.

11.2.6   Signal Converters

It is frequently necessary to modify the output of a transducer to satisfy the input requirements of indicators, recorders, displays, data readouts, and control elements or circuitry. Signal converters are employed to provide the required modification. Commonly used converters are divided, by function, into the following categories:

1. Modulators.
2. ac voltage to dc voltage.
3. ac current to dc voltage.

Table 11.3
Transducers for Mechanical Impedance and Dynamic Measurements

Application	Type	Typical Performance Characteristics			
		Sensitivity	Frequency (Hz)	Range	Impedance
General purpose for small and medium size specimens	Piezoelectric impedance head: Force	250 pC/lb	5–4,000	50 lb	600 pF
	Acceleration	8 pC/g	5–4,000	250 g	600 pF
	Piezoelectric: Force	45 pC/lb	2-4,000	500 lb	200 pF
	Wire strain gage: Force	200 mV/lb	0-100	0.1 lb	300 Ω
General purpose large-size specimens	Piezoelectric impedance head: Force	15 pC/lb	2-5,000	5,000 lb	5.7 nF
	Acceleration	165 pC/g	2-5,000	500 g	2.1 nF
	Piezoelectric: Force	15 pC/lb	2-5,000	5,000 lb	5.7 nF

Table 11.4
Transducers for Dynamic Pressure Measurements

Application	Type	Typical Performance Characteristics			
		Sensitivity (mV/psi)	Frequency (Hz)	Pressure (psi)	Impedance
General purpose for low and medium presures	Capacitive	3,000	10,000	5	—
	Piezoelectric	60*	9,000	500	300 pF
	Piezoresistive	5	1,000	100	1,700 Ω
	Wire strain gage	20	200	1	350 Ω
	Differential transformer	30	2,400 †	0.5	600 Ω
	Variable reluctance	10	3,000 †	100	1,000 Ω
General purpose for high pressures	Piezoelectric	1,000*	20,000	10,000	9,000 pF
	Piezoresistive	0.01	‡	10,000	700 Ω
	Wire strain gage	0.004	—	5,000	350 Ω
	Differential transformer	0.1	—	1,000	1,200 Ω
	Variable reluctance	0.02	3,000†	5,000	1,000 Ω
Blast studies	Piezoelectric	300*	100,000	500	2,100 pF

*Charge sensitivity, in pC/psi, can be obtained by multiplying the voltage sensitivity by the capacitance in manofarads.

†Carrier frequency, the transducer operates at lower frequencies

‡Function of excitation.

Source: "Survey of Transducers and Signal Converters", *Electronic Instrument Digest,* p. 37, Chicago: Milton S. Kiver Publications Inc., May 1969.

311

4. dc voltage to dc current.
5. dc current to dc voltage.
6. Voltage to frequency.
7. Frequency to dc voltage.
8. Frequency to frequency.
9. Phase to dc voltage.

Modulators may be of the electromechanical, magnetic, vacuum tube, or semiconductor type. Modulators are devices that mix two signals of different frequencies. Choppers provide high signal to noise ratios, have complete drive signal and output isolation, exhibit low input impedance, but are limited, generally, to applications in which signal frequencies do not exceed 2 kHz. Vacuum tube units have high input impedance and can be used at frequencies in excess of 10 kHz, but they are subject to considerable drift. Magnetic modulators are basically second-harmonic-operated saturable reactors that can achieve nulls on the order of 1 μV. Transistor switches, semiconductor diodes, photodiodes with intermittent light sources, and nonlinear resistors are also frequently employed as modulators. They permit operation over a wide range of frequencies and provide excellent stability and signal to noise ratios. Voltage-sensitive capacitors and voltage-variable capacitors may also be used as modulators.

Alternating voltage/current to dc voltage converters are available that provide wide application flexibility. ac inputs, from a few microvolts up to several hundred volts, can be accommodated over the frequency range from approximately 50 Hz up to one MHz. High input and output impedances are provided, and accuracies are of the order of 0.5%.

A variety of dc voltage to current and current to voltage converters are available, most of which use magnetic transistor-amplifier configurations. Some units of this type are referred to as *dc transformers*. Inputs in the ranges from 2 to 50 mV and from 0 to 3 A can be converted into currents ranging from 1 to 50 mA or into voltages up to 30 V.

Voltage to frequency converters accept dc inputs up to about 30 V and supply pulses at frequencies up to 250 kHz. Output frequency is proportional to input voltage, with a nonlinearity of less than 0.02%. Voltage to frequency converters also find application in analog to digital conversion as well as in data transmission. They are also used to send a transducer signal over 100 ft, as shown in Fig. 11.15.

Solid-state circuitry is generally employed to perform frequency to dc voltage conversion. Frequencies up to approximately 100 kHz provide linear outputs up to about 10 V. The units are capable of handling sine, square, and triangular waveform inputs as well as pulses.

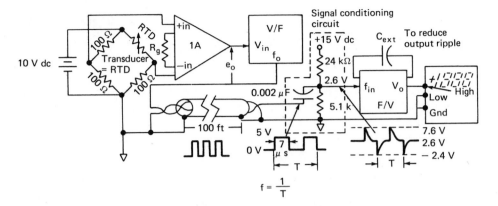

Figure 11.15 Transducer bridge with one leg designated as RDT. A resistor thermometer detector transducer is connected to an instrumentation bridge called 1A, which amplifies the signal from the transducer 1,000 times. This is connected to a voltage to frequency counter, V/F. A frequency to voltage converter, F/V, is added to the circuit and represents the analog signal to the digital panel meter. This F/V converter can make analog control signals available for other applications. Courtesy of E. L. Murphy, "Sending Transducer Signals Over 100 Feet?," *Instruments and Control Systems Magazine*, Vol. 49, June 1976, pp. 35–39.

Typical frequency to frequency converters change frequencies on the order of 1–10 kHz to lower frequencies, such as 400 or 800 Hz. These units are usually dc-powered, although they are also available with built-in power supplies.

Solid-state units for converting phase to dc voltage, with accuracies of $\pm 0.1°$, operate in the frequency range from 360 Hz to 10 kHz. They are capable of providing output over the full 360° range.

11.3   SOUND MEASUREMENTS*

11.3.1   Introduction

Various transducers are available for measuring sound in gaseous, liquid, or solid media. These transducers measure pressure gradient, which is directly related to particle velocity, with no general-purpose transducers available for measuring sound intensity or particle displacement.

Transducers can make sound pressure level measurements from approx-

---

*Section 11.3 appears in G. F. Harvey, ed., *ISA Transducer Compendium*, Part 1. Plenum, New York, 1969.

imately -30 to 250 dB, relative to 0.0002 μbar (10^{-10} to 10^4 psi in the frequency range from 0 to above 10^7 Hz).

11.3.2   Units and Conversion

A reference of 1 μbar is commonly used for underwater acoustics. A reference of 0.0002 μbar is most commonly used in air. A microbar is approximately 10^{-6} atmosphere. It is equal to 0.1 N/m² or 1.0 dyn/cm². Standard sound level meters are calibrated in decibels relative to 0.0002 μbar.

11.4   SOUND TRANSDUCERS

Microphone transducers are generally classified according to the following names:

1. Condenser microphones.*
2. Moving coil or dynamic microphones.
3. Ribbon microphones.
4. Ceramic microphones including crystal type.
5. Magnetostrictive transducers.
6. Hot wire probes.
7. Piezoresistive microphones.

11.4.1   Calibration Measuring Techniques

Calibration measuring techniques [7] to be made on microphones include

1. Reciprocity calibration.
2. Calibration by substitution.
3. Pistonphone.
4. Electrostatic actuator.
5. Selection, placement, and orientation of microphones are important in measurements.

11.4.2   Microphone Sensitivity Relationships

A sensitivity of -60 dB re 1 V/μbar equals 6.85 V/psi, since 1 μbar equals 1 dyn/cm² and 1 dyn/cm² equals 14.7 psi.

A sensitivity of -60 dB re 1 V/μbar also equals 1 mV/dyn/cm², and equals 6.9 V/psi which equals $\frac{4}{3}$ V/mm Hg.

---

*Condenser cartridges generate a voltage because the voltage across a capacitor having a charge is proportional to the distance between the plates.

11.5 THE MINIATURE (CERAMIC) ELECTRET MICROPHONE*

In view of the requirements for improved miniature transducers for acoustic measurements, hearing aids, stereo phonograph cartridges, tweeter loud-speakers, communications, data transmission, etc., a study was carried out on the performance of small microphones. A series of miniature electret (ceramic) microphones has been developed for the specialized applications mentioned above. Up to now, electret microphones have been subject to humidity, temperature, and drift and have changed with time. In the electret microphone adaptable for medical applications, an electret foil diaphragm is periodically supported across the face of the backplate by raised areas projecting above the backplate surface, as shown in Fig. 11.16.

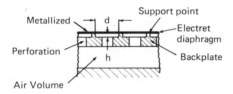

Figure 11.16 Multiple diaphragm support used in electret cartridge. From F. Frain and P. Murphy, "Miniature Electret Microphone," *JAES*, Vol. 18, No. 5, October 1970, p. 512.

11.6 VIBRATION TRANSDUCERS

Accelerometers, velocity coils, and proximity probes are all types of vibration transducers. The velocity of a point of a vibratory surface is the integral (area under curve) of its acceleration, and its displacement is an integral of the velocity. Accelerometers, whether piezoelectric, piezoresistive, servo, capacitive, or inductive, all act as spring/mass systems transforming the force acting on the mass into an analog electrical output. Each of these types of accelerometers has certain advantages for different application. Considerations in choosing the proper accelerometer are frequency range, acceleration range, shock levels, temperature, size, ruggedness, and the matching electronics required.

Just as a good mechanic can often tell when a machine is going bad by the way it sounds, so too, the use of accelerometers on a structure detects changes in the vibration level. Furthermore, analysis of the normalized amplitude of the vibrations vs. frequency gives much greater information

---

*Section 11.5 is based on information found in F. Fraim and P. Murphy, "Miniature Electret Microphones," *JAES*, *18*, No. 5, pp. 511–517 (Oct. 1970).

for analysis. These frequency response data are highly useful whether the subject is a jet engine, a power plant turbine, or a television set package drop test as used in television set quality control.

Underwater, sound can be measured by a sound level meter, provided a suitable hydrophone or accelerometer is used as the microphone source.

11.7   MECHANICAL IMPEDANCE MEASUREMENTS

Design testing and reliability testing of electronic equipment are carried out in a television quality-control production system, in space on a missile, or in a loudspeaker system where mechanical vibrations are important. In such studies, the absolute electrical impedance is the *ratio* of voltage to current, and the *mechanical* impedance is the ratio of force to velocity. Mechanical impedance testing [8] requires the following items for measurement:

1. The source of the vibratory force variable over the frequency range.
2. Method or methods of measuring the vibratory force applied to the specimen.
3. Method or methods of measuring the resultant motion of the force applied to the specimen.
4. Method or methods of measuring the phase angle between the velocity and force vectors or phasors.
5. Method or methods of plotting and recording the output as a function of frequency.

An electrodynamic vibration generator or loudspeaker shaker system to generate force can be used from about 5 to 10,000 Hz. The shaker is usually mounted on a large mass which will absorb reaction forces.

Force and motion measurements can be achieved by a ceramic gage and a piezoelectric accelerometer, respectively.

11.8   HIGH-FREQUENCY LOUDSPEAKER
TRANSDUCER

The high-frequency loudspeaker is the last link in the chain of components that make up the reproducing system of a high-fidelity system. It is the high-frequency loudspeaker, or *tweeter*, as it is called, which transforms the high-frequency electrical energy into sound. Transducers of this type generally consist of some kind of diaphragm which is vibrated by some mechanical force produced by electrical energy; between the diaphragm and the air, a coupler must be provided so that the acoustic energy will be radiated efficiently. The three types of tweeters used today are the direct radiator, the horn-loaded, and the electrostatic types. The direct radiator tweeter may be further subdivided into dynamic and ribbon types. If the tweeter is to reproduce the sound faithfully, it must create a pressure or particle velocity at its output which is an exact duplicate of the electrical waveform that appears at its input terminals. To achieve this, high compliance and low mass of the diaphragm, high flux density in the magnetic gap, smooth and extended response, nondirectional characteristics, and proper damping of the oscillatory system are all required. The high-frequency loudspeaker must also generate sound waves that are free of any nonlinear distortion. Measurement techniques have been discussed at length in Chapter 9 and are used to determine the dynamic performance of a high-frequency loudspeaker or tweeter.

11.9   NO-TOUCH SENSORS

Almost all no-touch controls involve transducers that produce an output voltage in response to a moving target. The target may be an opaque material (optoelectronic controls), a person (sonic alarms), a finger (capacitance), a metal (eddy current), or a magnet (Hall effect). See Table 11.5.

11.10   SUMMARY

We have examined various transducers with emphasis on the measurements that can be made with them. Some theory has been introduced. Insight into microphones and high-frequency loudspeakers is used to further illustrate the potential of transducers used in electronics. Each of the transducers can be examined in an experimental lab. Without transducers, electronics could not have attained the important role it plays in real-life situations. Our electronic instruments would be nothing more than laboratory or production phenomena. With transducers we can use electronic instrumentation to measure, modify, and improve the technological world we have created.

Integrated temperature transducers presently are available with output operational amplifiers. Such transducers have a sensitivity per degree centigrade of 10–200 mV and are very linear. They are available from the National Semiconductor Corp. of Santa Clara, California.

Ion-sensitive field-effect transistors are now used as transducers in measuring bioelectronic potentials.

One further use of transducer systems in electronic measurements is in the field of telemetry (an electronic instrument system for recording at a distance the readings of other instruments). In this system, a transducer is fed to a conversion system which is modified for telephone transmission. The transmitted signal is converted at the receiving end by a conversion scheme to an analog or digital signal.

Table 11.5
No-touch Sensors: How They Work

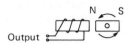

Output

Magnetic reluctance: The sensor is, in essence, a dc iron core inductor. When the target, a high-permeability metal pole piece, rotates past the inductor, the filed is interrupted and a voltage is produced.

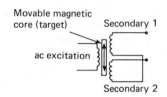

Movable magnetic core (target)
Secondary 1
ac excitation
Secondary 2

Variable reluctance: In these devices, the dc field is replaced by an ac field, and the output voltage varies as the impedance of the inductor is changed as the pole piece moves through the coil. The addition of two symmetrically spaced secondary coils makes it a linear variable differential transformer.

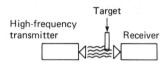

Target
High-frequency transmitter
Receiver

Sonic and ultrasonic: A high-frequency signal (usually above 16 kHz) is transmitted at one point and received by a transducer at another point. A target interrupting the signal path or reflecting a change in frequency can activate an alarm.

Sensor
Target

Capacitance: The sensor is the stationary plate of a capacitor, and the movable target serves as the other plate. As the target approaches the fixed plate, the change in dielectric constant produces a change in signal.

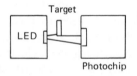

Target
LED
Photochip

Optoelectronic: These controls consist simply of a light source directed toward a photoreceiver. When the light path is blocked by the target, a change in signal occurs. Depending on the electronics, an external load can be either a low or a high-level signal.

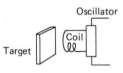

Oscillator
Coil
Target

Eddy current: Designed for detecting metals, these sensors consist of an oscillator that produces a field in a sensing coil. A metal target brought in front of the coil produces an opposing magnetic field that "kills" the oscillator and results in a signal change.

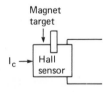

Magnet target
I_c
Hall sensor

Hall effect: These sensors are based on the fact that when a magnetic field is applied perpendicular to a fixed current flowing through a semiconductor material, a voltage is produced that is proportional to the flux density. In the Hall device, the change in flux is caused by movement of a permanent magnetic target.

Darrell West, "Touch-Me Not," *Electronic Products,* Jan. 1976, p. 41.

REVIEW QUESTIONS

1. What is a transducer?
2. Why are transducers important in electronic instrumentation?
3. List five transducers that can be used to measure pressure.
4. List five items that transducers measure.
5. What are eight important parameters for the design of a good transducer?
6. Describe a strain gage.
7. a. What is a semiconductor strain gage?
   b. Where is it used?
8. List three types of temperature transducers, and describe the uses of each.
9. Draw a circuit using a thermistor.
10. Explain how to use a potentiometric transducer.
11. Draw a linear differential transformer transducer.
12. Describe the operation of a piezoelectric transducer.
13. a. What is a signal converter?
    b. Give some examples.
14. What is a microbar?
15 List three types of microphones.
16 What is a ceramic transducer?
17. Explain the operation of the bimorph construction of a ceramic element.
18. What is a vibration transducer?
19. What is an electret microphone?
20. What is meant by mechanical impedance?

GENERAL REFERENCES

1. G. F. HARVERY, ed., *ISA Transducer Compendium*, Part 1, 2nd ed. Plenum, New York, 1969.
2. G. F. HARVERY, ed., *ISA Transducer Compedium*, Part 2, 2nd ed. Plenum, New York, 1969.
3. *Clevite Brush Displacement Transducer Bulletin 639-5*, Metripak Transducer Models 33 03, 33 04, 33 05, 33 06, July 1, 1967.
4. A. E. ROBERTSON, *Microphones*, 2nd ed. Hayden Book Co., Rochelle Park, N.J. 1963.
5. L. BERANEK, *Acoustics*. McGraw-Hill, New York, 1964.

6. F. Fraim and P. Murphy, "Miniature Electret Microphones," *JAES*, *18*, No. 5 (Oct. 1970), pp. 511-517

7. "Survey of Transducers and Signal Converters," *Electronic Instrument Digest* (May 1969), pp. 31–40. Published by Milton S. Kiver Publications Inc., Chicago, Ill.

8. *Introduction to Transducers for Instrumentation.* Statham Instruments, Inc., Oxnard, Calif., 1968.

9. Kenneth Arthur, *Transducer Measurements*, 2nd ed. Tektronix, Inc., Beaverton, Ore., 1971.

10. *Biophysical Measurements*, Measurement Concept Series. Tektronix, Inc., Beaverton, Ore, June 1973.

11. Edward E. Herceg, *Handbook of Measurement and Control.* Schaevitz Engineering Co., Pennsauden, N.J., 1976.

12. W. D. Cooper, *Electronic Instrumentation and Measurement Techniques*, 2nd ed. Prentice Hall, Englewood Cliffs, N.J., 1978.

REFERENCES

1. "Strain Recording Techniques," *Data Display*, Vol. 1, No. 3. Gould Inc., Instrument Systems Division, Cleveland.

2. "Taking Industry Temperature," *Data Display*, Vol. 1, No. 2. Gould, Inc., Instrument Systems Division, Cleveland.

3. *YSI Precision Thermistor Catalogue.* Yellow Springs Instrument Co., Yellow Springs, Ohio, Oct. 1974.

4. L. C. Lynnworth, "Industrial Applications of Ultrasound—A Review. II. Measurements, Tests and Process Control Using Low-Intensity Ultrasound," *IEEE Trans. Sonics and Ultrasonics, SU-22*, No. 2 (March 1975), pp. 71-109.

5. "Meet the Metrisite, Our Contactless Angular Position Transducer," *Data Display*, Vol. 2, No. 4. Gould Inc., Instrument Systems Division, Cleveland.

6. *Gould Biophysical Instrumentation, Catalog 998-14A.* Gould Inc., Instrument Systems Division, Cleveland, Dec. 1974.

7. *Acoustics Handbook, Application Note 100.* Hewlett-Packard, Palo Alto, Calif., Nov. 1968.

8. Wilson Bradley, Jr., "Mechanical Impedance Testing," Endevco Tech. Paper, TP 202. Endevco Corp., San Juan Capistrano, Calif., 1964, pp. 1–10.

12

SIGNAL GENERATION: APPLICATIONS AND MEASUREMENTS

12.1 INTRODUCTION

In the field of electronics, there are two types of electrical signals (voltages or currents) which are extensively used. One type of signal is the sinusoidal waveform which is utilized in high fidelity (Chapter 9) and is also extensively used in the transmission and reception of information (TV, radio). The other type of signal commonly used in electronics is nonsinusoidal waveforms such as pulses, square waves, and ramp waveforms. In this chapter we shall discuss the basic elements of instruments which produce these signals and their relation to basic electronic measurements.

Devices which produce these signals for the purpose of testing electronic circuits and systems are called generators. In the material which follows we shall cover function generators, oscillators, and pulse and square-wave generators. It should be emphasized here that distortion-free sine-wave output can be checked by an oscilloscope (Chapter 5) and by spectrum analyzers (Chapters 9 and 13). The important instrument utilized in measuring frequency, the electronic counter, will also be discussed in detail here.

12.2 FUNCTION GENERATORS—CONCEPTUAL REMARKS

The function generator is an important signal source because it provides a variety of wave shapes, including triangular, square, sine, and pulses. Triangular waveforms, used in conjunction with oscilloscopes, can determine the overload (clipping) point of amplifiers. Square waveforms simultaneously reveal low-frequency response (by sag), high-frequency response (by rise time), and transient response (by ringing and other aberrations) of amplifiers. Sine waves, universal in the electronics industry, show the full frequency response of various devices. Further, pulses and square waves are used as clock and signal sources in logic circuitry. Ramps and triangles provide time bases for oscilloscopes and paper recorders and test signals for voltage comparators. The high-frequency range of modern function generators extends from the audio into the video-radio spectrum and is useful in the telecommunications field as a modulation signal source. Their extreme low-frequency range is useful in biological and geophysical simulations, servo systems, mechanical testing and simulations, and other applications. Material from this section appears in the 1976 *Tektronix Product Catalogue.*

12.3   OSCILLATORS

12.3.1   Conceptual Remarks

Alternating current signal sources are described by various names: oscillators, test oscillators, audio signal generators, etc. Different names are applied depending on the design and intended use of the source. The oscillator is basic to all the sources and generates sine-wave signals of known frequency and amplitude. In the recently developed transistorized sources, the name *test oscillator* has been used to describe an oscillator having a calibrated attenuator and an output monitor. The term *signal generator* is reserved for an oscillator with modulation capability.

Function generators, which we have discussed in Section 12.2, have both sinusoidal and nonsinusoidal outputs.

12.3.2   Basic Oscillator Requirements

In selecting an oscillator, the technician will be most interested in its frequency coverage. The question to be answered here is, "Will the instrument supply both the lowest and highest frequencies of interest for anticipated tests?" Frequencies of 0.00005 Hz to over several hundred megahertz are possible.

Power output, oscillator's stability, dial resolution, harmonic distortion, hum, and noise in the output signal are also important specifications.

12.3.3   Testing Receivers Using
a Swept FM Signal Generator*

In this section we shall concentrate on the primary use of signal generators which are high-frequency sinusoidal sources with modulation capabilities for

*Section 12.3.3 is based on Hewlett-Packard's *Signal Generator Section Seminar* Sept. 1974.

receiver testing. There are many different types of receivers and we cannot discuss each one in detail, so we have chosen as a common denominator a frequency-modulated land mobile receiver.

With reference to the simplified block diagram of an FM receiver (Fig. 12.1), the first stage is the tuned RF section which in theory lets only one

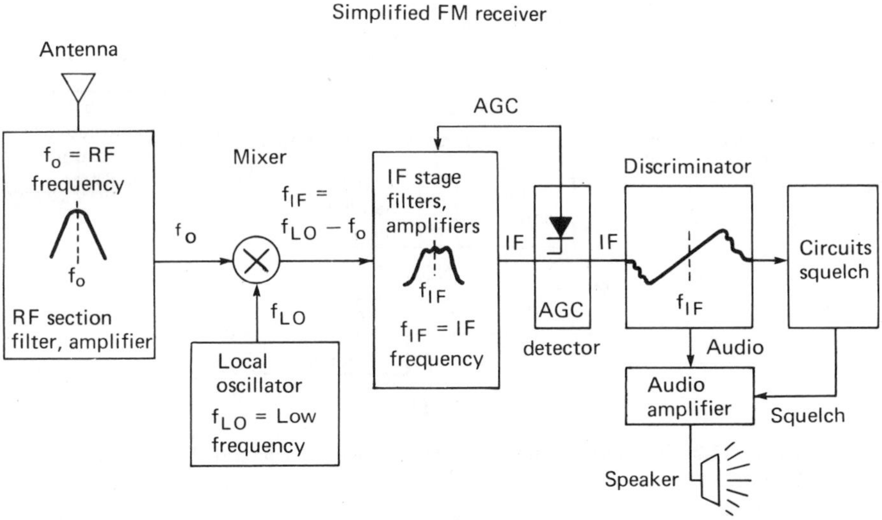

Figure 12.1 Simplified FM receiver. Courtesy of Hewlett-Packard, Palo Alto, Calif.

station at a time into the mixer of the receiver. Actually the RF section is coarsely tuned, with the fine tuning accomplished by the narrow bandwidth of the IF section. The local oscillator heterodynes the incoming RF to an IF signal into the mixer stage. The IF amplifiers are followed by an AGC detector, which controls the gain of one of the IF amplifiers. The signal then goes through a discriminator, or FM detector, where the audio is demodulated and fed to the audio amplifier and speaker. The squelch circuits monitor the audio, and when no audio is present they cut off the audio amplifier so the user does not have to listen to noise.

12.3.3.1 RF generator receiver coupling. Initially we shall explain how to connect the signal generator to the receiver. The purpose of all these tests is to use the signal generator to simulate a signal received out of free space by the antenna of the receiver. A measuring instrument, usually a distortion analyzer or oscilloscope, is then connected to the audio output of the receiver to measure the quality of the received signal (Fig. 12.2). It is important to couple the generator to the receiver in a manner such that we have calibrated output from the signal generator and an impedance is presented to the

Receiver testing

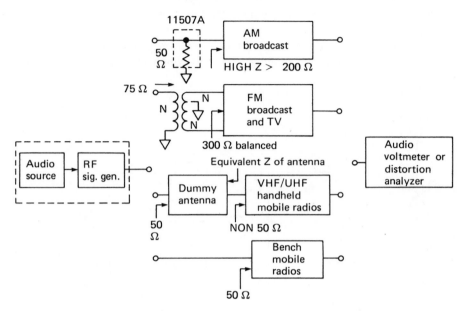

Figure 12.2   Receiver testing. Courtesy of Hewlett-Packard, Palo Alto, Calif.

receiver's antenna input that looks like that of the normal antenna. Dummy antennas are usually specified in the receiver test information. One type that is common is a 50-Ω resistor to ground used with a high-impedance (> 500 Ω) receiver. For FM broadcast and TV receivers with balanced antenna inputs, coupling is done with a balun or a resistive divider to give a balanced output. If the receiver has some other impedance, a complex dummy antenna is used. FM land mobile receivers usually have 50-Ω antenna inputs so the signal generator can be directly coupled to them.

12.3.3.2   Receiver alignment.   The IF filter/discriminator alignment is not usually a receiver test but rather is a procedure done in production testing and after some repairs (Fig. 12.3).

An FM swept generator is connected to the receiver and its RF is tuned away from a station. The receiver is tuned to the same frequency as the generator. The generator is then FM'ed internally at a very low rate, say 40–50 Hz (this can be done in an HP 8640 with the optional built-in variable frequency oscillator). The maximum frequency deviation is set much greater than the IF bandwidth of the receiver. The FM output is also used to exter-

IF filter discriminator alignment

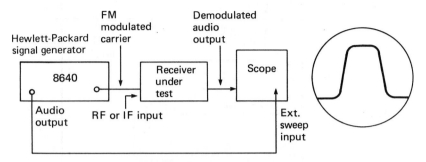

Test procedure: 1.   Tune receiver away from station and set 8640 at this
frequency.

2.   Set maximum deviation >> IF bandwidth of receiver
and FM output to external sweep input on scope.

Important sig. gen. specs: Wide deviation
Linear FM

Figure 12.3   IF filter/discriminator alignment procedure. The 8640 is
a Hewlett-Packard general-purpose signal generator. Courtesy of Hewlett-
Packard, Palo Alto, Calif.

nally sweep the scope. If the optional internal oscillator is not available,
then an external sawtooth can be used to externally FM the generator and
sweep the scope.

It is important that the generator have provision for wide enough FM
deviation so that it can be used as a sweeper for IF and discriminator align-
ments. Also, the FM should be linear to ensure that the only distortions
seen come from the receiver and are not introduced to the receiver by the
generator.

**12.3.3.3   Receiver response alignment using a spectrum analyzer-external RF
sweep generator.**   A special technique using a signal generator with a
Hewlett-Packard audio spectrum analyzer or equivalent to look at the total
audio response of a receiver is shown in Fig. 12.4. The tracking output of the
audio frequency spectrum analyzer (which tracks the analyzer input frequency) is used to externally modulate the signal generator. The modulated
RF output is applied to the receiver's antenna, with its audio output applied
to the input of the spectrum analyzer. This gives the swept frequency
response of the total receiver. This is particularly useful for analyzing the
audio circuits of receivers with multiple channels on the same carrier (e.g.,
multichannel SSB receivers).

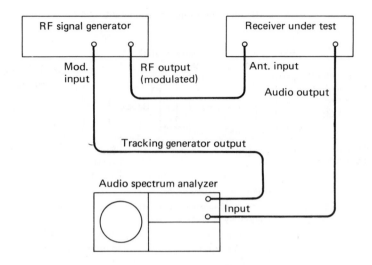

Figure 12.4 Test setup for receiver response. Courtesy of Hewlett-Packard, Palo Alto, Calif.

12.4 PULSE AND SQUARE-WAVE GENERATORS

12.4.1 Conceptual Remarks*

Electronic instruments such as pulse generators have come into play since 1940. With the advent of digital circuitry, the performance, quality, and reliability of such devices have improved. The pulse generator is an electronic instrument that gives an output which represents voltage and current and is visualized as a pulse or series of pulses (pulse train). The square wave is a pulse or series of pulses which have equal *on time* and *off time*. Pulse and square-wave generators most often are used with an oscilloscope as the readout device, which provides both qualitative and quantitative evaluations of the performance of the system or device under test.

The pulse generator finds itself in many applications because of the great flexibility of its output. Some of these applications are

1. Transient analysis of linear systems such as overload recovery of amplifiers and damping of servo systems.

2. Radar system testing where the pulse generator is used to simulate the video of a radar return. In these applications the variable delay of the pulser is used for range calibration and the amplitude is varied to test the sensitivity of the video circuits.

*Sections 12.4.1 through 12.4.7 are based on material from Hewlett-Packard, Palo Alto, California.

3. Component testing such as transistor recovery time and photomultiplier switching speeds.

4. Communication system testing. This application area is very varied. The pulse generator can be used to perform sensitivity tests like those of the radar example plus worst-case data generation and bit error rate measurements.

5. Integrated circuit testing. See Chapter 8 for a complete analysis of integrated circuit testing.

12.4.2 Pulse Characteristics

The six pulse parameters which determine the pulse characteristics shown in Fig. 12.5 are

1. *Base line:* Reference dc level at which pulse starts.
2. *Amplitude:* Voltage level from base line that pulse switches to:
    a. Polarity: Direction of pulse amplitude with respect to the base line, $+$ or $-$.
    b. Offset: Amount of dc that reference level is shifted from ground, $+$ or $-$.
3. *Transition time:* Time for pulse to switch from one level to another.
    a. Leading edge transition time: Usually measured from 10% to 90% of pulse amplitude.
    b. Trailing edge transition time: Measured from 90% of pulse amplitude to 10% of pulse amplitude.
4. *Width:* Measured in seconds or fraction of seconds.
    a. The fixed transition time generator is measured from 50% point on leading edge to 50% point on trailing edge.
    b. The variable transition time generator is measured from initial corner of leading edge to initial corner of trailing edge.
5. *Repetition rate:* How often pulses occur.
    a. Rate: Measured in hertz.
    b. Period = 1/rate: Measured in time is the point on one pulse in the train to the corresponding point on the next pulse in train.
6. *Duty cycle:* Ratio of pulse width to pulse period expressed as a percent. If the period is 1 s and the pulse width is 0.1 s, then the duty cycle is $(0.1/1.0) \times 100\%$ or 10%.

Additional parameters required to describe actual pulses and pulse generators, described in Fig. 12.6, are

1. *Preshoot:* Amplitude perturbation immediately preceding transition, expressed as a percent of pulse amplitude.
2. *Overshoot and rounding:* Amplitude perturbation immediately following transition time, expressed as a percent of pulse amplitude.
3. *Ringing:* Amplitude perturbation following overshoot, usually a damped sinusoid, expressed as a percent of pulse amplitude.
4. *Settling time:* Time for overshoot and rounding to be within a specified percent of pulse amplitude.
5. *Nonlinearity:* Vertical variation in transition from straight line through 10–90% points, expressed as a percent of amplitude.
6. *Droop:* Decrease in pulse amplitude with time.
7. *Jitter:* Uncertainty in turning off pulse.

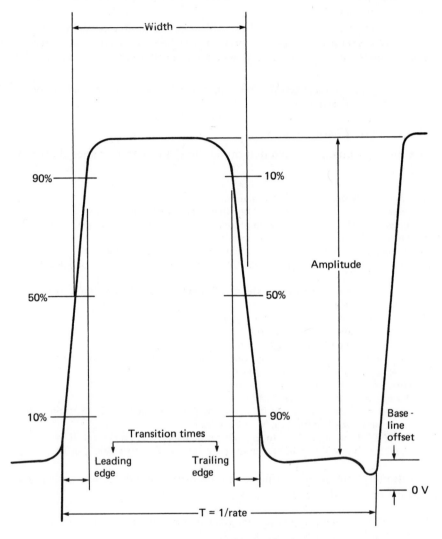

Figure 12.5   Pulse specifications.

a. Width: Uncertainty regarding turning off pulse.
b. Period: Uncertainty regarding when pulse turns on.

12.4.3   *Square Waves or Pulses*

The fundamental difference between pulse and square-wave generators concerns the signal duty cycle. Square-wave generators have equal *on* and *off* periods, this equality being retained as the repetition frequency is varied.

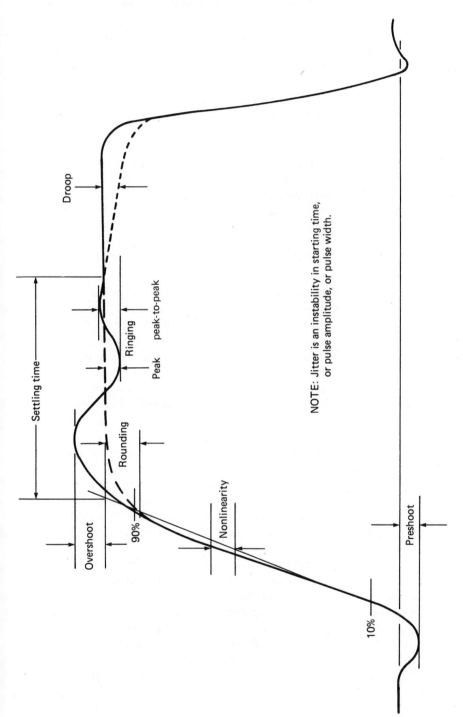

NOTE: Jitter is an instability in starting time, or pulse amplitude, or pulse width.

Figure 12.6 Real pulse.

329

The duration of a pulse generator *on* period, on the other hand, is independent of the pulse repetition rate. The duty cycle of a pulse generator can be made quite low such that the instrument can supply more peak power during the on period than square-wave generators.

Short pulses reduce power dissipation in the component or system under test. For example, measurements of transistor gain are made with pulses short enough to prevent junction heating and the consequent effect of heat on transistor gain.

Square-wave generators are used where the low-frequency characteristics of a system are important, such as in the testing of audio systems. Square waves also are preferable to short pulses if the transient response of a system requires some time to settle down. Refer to Chapter 5 [Sect. 5.18] for a discussion of the square-wave testing of an amplifier.

12.4.4   Selection of a Pulse Generator

In the selection of a pulse generator, the quality of the output pulse is of primary importance. High-quality test pulses ensure that degradation of the displayed pulse may be attributed to the test circuit alone.

The pertinent characteristics of a test pulse, shown in Fig. 12.5, are controlled and specified accurately in pulse generators. Rise and fall times should be significantly faster than the circuits or systems to be tested.

Any overshoot, ringing, and sag in the test pulse should be known, so as not to be confused with similar phenomena caused by the test circuit.

The range of pulse-width control should be broad enough to fully explore the range of operation of a circuit. Narrow pulse widths are useful in determining the minimum trigger energy required by some circuits.

Maximum pulse amplitude is of prime concern if appreciable input power is required by the tested circuit, such as a magnetic core memory. At the same time, the attenuation range should be broad enough to prevent overdriving the test circuits as well as to simulate actual circuit operating conditions.

The range of pulse repetition rates is of concern if the tested circuits can operate only within a certain range of pulse rates or if a variation in the rate is needed.

Pulse generators with fast rise times are widely used in the development of digital circuitry. Teamed with a suitably fast oscilloscope, these generators enable evaluation of transistor and diode switching times.

Variable rise-time and fall-time pulses are invaluable for testing devices whose output changes with rise time and fall time, such as magnetic memories. Variable transition time pulses are useful in checking logic circuitry where the input signal characteristics must be carefully specified.

Pulse generators are used as modulators for klystrons and other RF sources to obtain high peak power while maintaining low average power.

12.4.5   Pulse Generator Block Diagram

The instrument that is capable of generating a pulse is considerably more complex than the sine-wave generators. Only two variables are required to completely describe a sine wave, frequency and amplitude. A pulse, on the other hand, requires a minimum of eight variables to describe it.

To understand the operation of this complex series of instruments, let us look at a block diagram of a typical pulse generator. The pulse generator is made up of three basic blocks, rate, timing, and output shaping.

The block diagram in Fig. 12.7 gives the relative position of each major function and control of a pulse generator. Figure 12.8 explains the use of the various operating modes of these instruments.

We shall now describe the basic pulse generator functions.

Three basic function blocks of the pulse generator provide control of the various pulse parameters.

1. The rate generator determines when pulses happen in four ways:
   a. Internal: Set by front-panel control.
   b. Externally triggered: To match external rate.
   c. Single pulse: Manual control.
   d. Gating produces pulse bursts:
      (1) Synchronous: Pulses turn on fixed time after gate and complete last pulse.
      (2) Asynchronous: Pulses free-run when gate signal turns on, and the last pulse may not be completed when the gate signal turns off.
2. The timing block is used to set up special timing within the train or to notify the outside world of events taking place inside the pulse generator.
   a. Delay of drive pulse with respect to trigger.
      (1) Fixed delay.
      (2) Variable delay.
   b. Double pulse: Produces two pulses for one rate pulse.
      (1) Trigger and first pulse coincident.
      (2) Second pulse variable in time.
   c. Digital timing set with the word generator.
3. Shaping controls (nontiming parameters).
   a. Width.
      (1) Internal set by front-panel control.
      (2) External set by threshold crossings.
   b. Transition time.
      (1) No control in fixed transition time generator.
      (2) Variable generator time varied by front-panel control.
   c. Polarity either + or −.

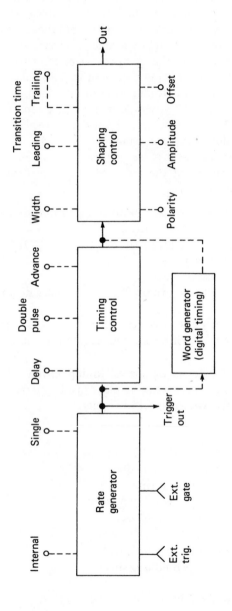

Figure 12.7 A basic pulse generator.

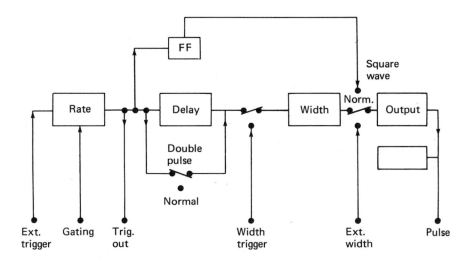

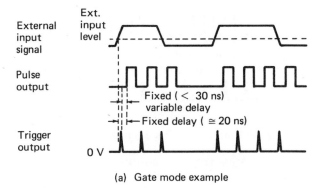

(a) Gate mode example

Figure 12.8 Pulse generator operational modes. a) Gated mode: When the pulse generator is used to perform tests of devices that are sensitive to a specific number of pulses such as a shift register, this mode of operation is required. The output is enabled during the time the gate signal is above the trigger level. All the normal front panel controls are active in this mode. b) External trigger mode: this is the second most common mode of operation of a pulse generator. In this mode the period of the output is controlled by the trigger input and all of the front panel controls are operational. When operated in the external trigger mode the pulse generator can be synchronized to other signal sources.

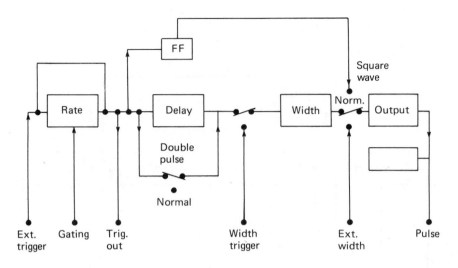

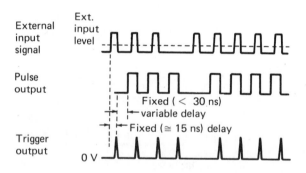

(b) External trigger mode example

Figure 12.8   (*Continued*)

d. Amplitude.
e. Offset either $+$ or $-$.

12.4.6   *Termination of a Pulse Generator into a 50-Ω Output*

With the output of a pulse generator consisting of pulses with a rise time on the order of 1 ns, its frequency spectrum output is equivalent to a sinusoidal signal operating at a frequency determined by

$$f_h = \frac{0.35}{t_r} = \frac{0.35}{10^{-9}} = 350 \text{ MHz} \qquad (12.1)$$

At this frequency, the pulse generator must be properly terminated with its characteristic impedance of 50 Ω to prevent reflections.

12.4.7   Output Format of a Pulse Generator

Because the pulse generator is widely used in digital circuit testing, it must be able to produce signals that are in a format that is acceptable to digital systems. For every specialized system a special pulse generator, known as a *word generator*, is used that has its timing block (Fig. 12.7) replaced by a data generator. Usually word generators do not offer all the shaping controls that are found on a pulse generator.

By adjustment of various controls, the pulse generator output can be a square wave which is a pulse output with a duty cycle of 50%. This is a very useful *setup* mode for the pulse generator because the width *and* delay controls are disabled. All the output shaping controls such as transition time, offset, and amplitude are active.

See Fig. 12.9 for a duty cycle that can be varied over very wide limits in the pulse mode. Also see Fig. 12.10 for a complement, symmetrical, and offset pulse.

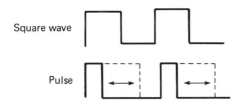

Figure 12.9   Duty cycle can be varied over very wide limits in the pulse mode.

12.4.8   Using a Pulse Generator as a Testing Device

Besides all of the applications mentioned in Section 12.4.1 the pulse generator can be used with a specialized oscilloscope known as a synchroscope which is used in radar to detect faults on transmission cable [1, 2] or telephone lines. A pulse travels down a transmission line at the speed of light, and when it detects an open line it is reflected toward the generator. At the generator end, the initial and reflected pulses are detected. The line length at which the fault occurs is found by,

$$\text{line length in miles} = \frac{186,000}{2} \times t_d \qquad (12.2)$$

where t_d is the delay time between the generated and reflected pulse. See Fig. 12.11.

The synchroscope sweep is calibrated in feet or miles per division and by simply detecting the delay of the reflected pulse, telephone line faults may be

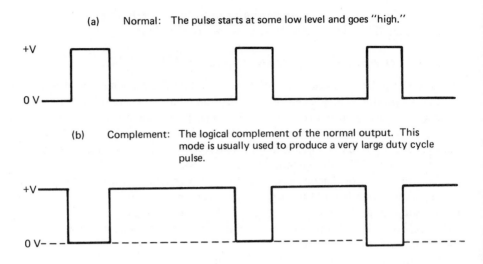

(a)   Normal: The pulse starts at some low level and goes "high."

(b)   Complement: The logical complement of the normal output. This mode is usually used to produce a very large duty cycle pulse.

(c)   Symmetrical: The pulse amplitude is symmetrical about the 0 = level.

(d)   Offset: A dc shift in the 0-V level of all the above outputs.

Figure 12.10 Complement, symmetrical, and offset pulse.

easily determined. Such reflection techniques are also discussed in Section 13.6.

12.5 BASIC ELECTRONIC COUNTER PRINCIPLES*

12.5.1 *Introduction*

In a digital computer, a number can represent an instruction as well as physical quantities. Of prime importance are electronic circuits that count. The transduction process converts electrical, mechanical, or optical information into electrical impulses that can be counted and controlled. Some of

*Parts of Section 12.5 are based on and excerpted from Ray Herzog, "A Review of Basic Counter Principles," *Tekscope*, March 1971, copyright Tektronix, Inc., Beaverton, Oregon.

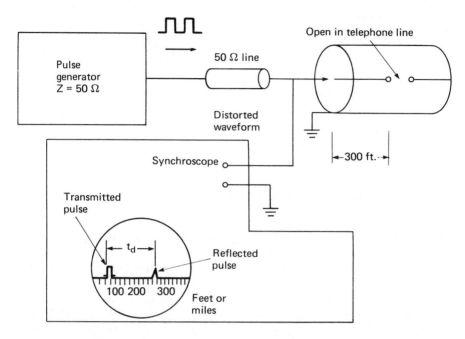

Figure 12.11   Pulse generator test setup, for finding an open on a telephone line. The synchroscope used is a radar oscilloscope.

the conversion devices are photocells, magnetic coils, switches, and transducers that measure pressure, temperature, velocity, acceleration, and displacement.

12.5.2   Basic Counter Functions

Although there are many counter types, all are basically designed to measure an unknown event, it being frequency or a time period. The direct counting type of counter consists of five main functions essential to any counter:

1. Signal input conditioner.
2. Gate.
3. Gate control/time base.
4. Decimal counting units (DCUs).
5. Readout device.

See Fig. 12.12.

The signal conditioner is a signal processing circuit which increases or reduces the amplitude of the incoming signal. Important criteria for such a device include selection of coupling and impedance matching. This amplifier

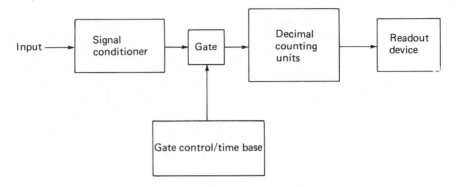

Figure 12.12   Five main functions of an electronic counter.

also transforms the measured signal's waveform into a precisely shaped signal suitable for further counter functions. The conditioner may also be called a *shaper*. The need for signal shaping as in the case of the recording process arises from the fact that input signals, with varying shapes and amplitudes, are not suitable to drive the counting circuit.

Signal shaping is usually done with a Schmitt trigger circuit, which is a type of multivibrator. Inherent in this Schmitt trigger is a level/slope control that selects the amplitude and slope on the signal where the counter is triggered.

The conditioned signal is next passed through a gate for a time interval determined by another function: the gate control/time base. The gate is a "go/no-go" device and determines the length of time the conditioned signal is fed into the decimal counting unit.

Signal pulses from the gate, having been determined by the counting mode, are fed to the decimal counting units (DCUs) where they are converted into a signal suitable to drive the readout device. DCUs are usually flip-flops arranged to divide their input by 10; they drive the readout in binary-coded decimal (BCD) form. The first DCU gives the "units" count; the second DCU, the "tens" count; and so forth. The number of DCUs, as well as the readout capacity, determines the magnitude of the displayed count. As an illustration, eight DCUs and a corresponding number of readout units give an eight-digit readout.

The readout devices provide a visual indication of the count number. Typical readout divices include neon lamps, incandescent lamps, light-emitting diodes, gas ionization tubes, and multisegment bar indicators. We can obtain an oscilloscope readout with an alphanumeric display of information on the CRT on a time-sharing basis along with the analog waveform. Such an oscilloscope has been used with the Tektronix 7D14 counter.

12.5.3 Modes of Counter Operation

The electronic counter is most often thought of as a device that totalizes (counts) input devices. This operation in the totalize mode is only one of the seven common modes. A counter can also indicate an input's frequency or period in frequency and period modes. It can compare two signals in the ratio mode. It can indicate the time between two points on a waveform; when they represent an input's pulse width, the counter would be in a width mode.

Finally, in the averaging mode, a counter can average the measurement reading over a number of periods or time intervals to give better resolution. We should realize that not all counters are capable of seven modes.

12.5.3.1 Totalize mode. In this mode the gate control/time interval is a simple switch that turns the gate on and off in performing a totalize operation. In totalizing the input signal, all that is required is to let the signal accumulate the readout register. In this case, the gate is permitted to pass the signal for whatever time the totalizing is to occur.

In Fig. 12.13, the totalize mode, the gate is turned on for the time that the input signal is to be accumulated.

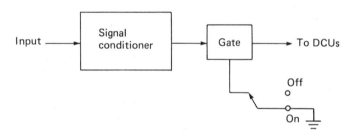

Figure 12.13 Totalize mode.

With the gate control set to external, the time for totalizing can be manually controlled by a switch shown in Fig. 12.14.

In both of the above totalizing operations, the input signal will be totalized (counted) for as long as the gate is conducting.

12.5.3.2 Frequency mode. When the gate is controlled by an accurate time interval the counter is in the frequency mode. This is shown in the block diagram in Fig. 12.15. The frequency mode is similar to the totalize mode since the input signal is counted for the period of time that the gate is open. The only difference between the totalize mode and the frequency mode is the way the gate control is operated.

For frequency measurements, the counting time interval must be accurate. Usually an internal crystal-controlled reference oscillator provides the

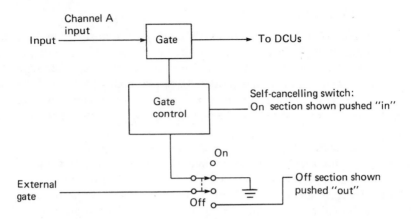

Figure 12.14 Two methods for totalize mode gate control: internal (manual) and external.

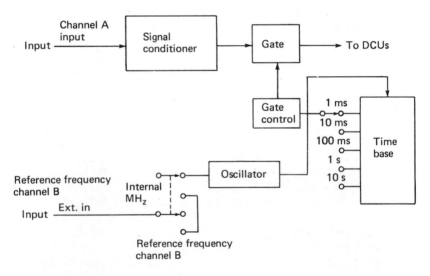

Figure 12.15 Two methods for frequency mode gate control: internal time base and external.

time interval. Some counters have provisions for supplying the reference oscillator externally.

12.5.3.3 Ratio mode.   A third mode of operation is the ratio mode (Fig. 12.16) in which the ratio of two input signals is displayed on the counter readout. The higher-frequency signal is fed to the channel A input connector and goes to the gate. The lower frequency signal is fed to the "ref. freq./ch. B ext. in" connector. With the ext. in switch set to ratio, this signal

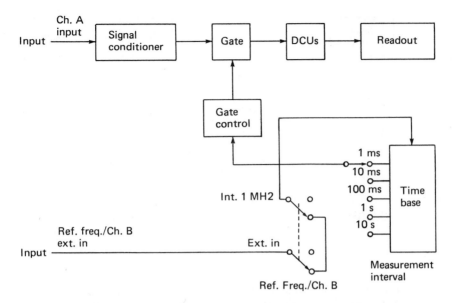

Figure 12.16   Ratio mode gives the ratio of two input signals.

is then routed to the time base. Here it is divided down and serves as the gate control signal.

In effect, what happens is that the lower-frequency signal determines how long the gate is open and thus how long the gate passes the higher-frequency signal from the channel A input. The gate output, therefore, is the ratio of the two signals.

12.5.3.4   Period mode.   The period of a signal is the reciprocal of its frequency. As such, the measure of the signal period might be expected to have a similar inverse relationship with its frequency measurement. And as shown in Fig. 12.17, the period mode circuit is similar to that of the frequency mode (Fig. 12.15) with the exception of a reversal of the gate inputs.

In the period mode, the signal to be measured is fed to the gate control where it determines how long the gate is open. An accurate oscillator signal is also fed to the gate. The gate output, therefore, consists of pulses from the oscillator that represent the measured signal's period. For example, say the signal to be measured is 100 kHz (10 ms) and the reference oscillator is 1 MHz (1 ms). The gate would be open for 10 ms, and in this time 10 pulses would pass from the oscillator. These 10 pulses are then processed by the counter to provide a readout indicating a period of 10 ms.

12.5.3.5   Time interval mode.   The time interval mode permits the counting of any number of events occurring between any two points on a waveform. This variable interval of measurement time is possible with a more elaborate gate than with other modes previously discussed. As shown in Fig. 12.18,

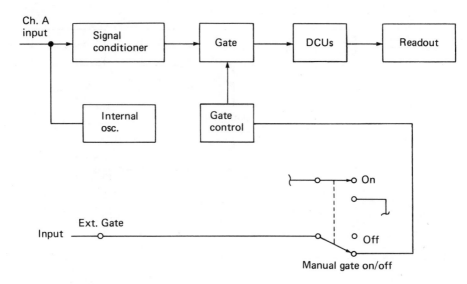

Figure 12.17 In period mode, counter counts a reference oscillator signal during an input signal's period.

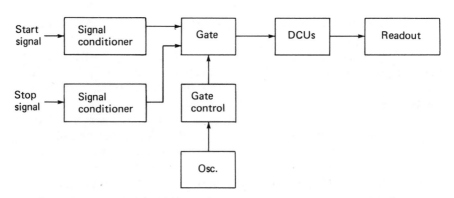

Figure 12.18 Start and stop signals open and close the gate in the time interval mode.

the gate has three inputs: one from the gate control and two from the signal conditioner stages, *start* and *stop* inputs.

The gate control feeds an accurate oscillator frequency to the gate where it then gets counted for a time determined by the start and stop input. The start and stop points are selected by triggering levels on the input waveform.

In a way, the time interval mode is like the period mode; i.e., counting is done during a given time. But with the selectable start and stop points, the time interval mode can count for not only the signal's period but for less than or more than a given period.

12.5.3.6 Width mode. The width mode is a type of time interval mode wherein the measurement is that of the signal's width. Compared with the time interval mode, the width mode would have a preset trigger and other circuitry to function on the signal slopes so that the effective start and stop points are those of the successive rising and falling slopes of the measured signal.

12.5.3.7 Averaging mode. It is often desirable when making period, time interval, or width measurements to be able to average the reading over a number of periods or time intervals to achieve better resolution and accuracy. This is done with the averaging mode.

Consider the period mode given in Fig. 12.17: If the input signal fed to the ext. gate connector were to be devided by 1,000, then the count of the reference oscillator applied to the channel A input would be averaged for 1,000 periods. As long as the total count does not exceed the capacity of the readouts, the resolution would effectively be increased 1,000 times.

In a ratio measurement, averaging is done when the signal fed to "ch. B ext. in" is routed through the time base to the gate control. In the ratio mode example, the 1-kHz input to the time base would be extended 1,000 times with the measurement interval switch in the 1-ms position (1 ms being 1/1,000th of a second).

12.5.4 Prescaling Counter

The prescaling counter (shown in Fig. 12.19) divides the input signal before it goes to the gate for subsequent counting. This design is economical; and

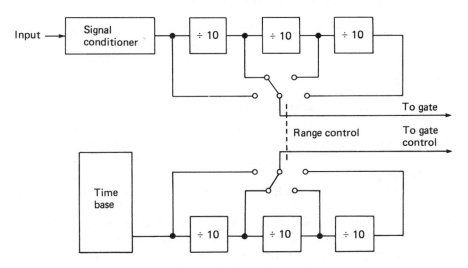

Figure 12.19 Prescaling counter divides input signal and time base before further processing and counting.

with a frequency range from dc to 1 GHz, this type of counter is popular.

On the other side of the pros and cons for prescalers are such things as not being able to count a number smaller than the divider ratio, reduced resolution, and longer time required to perform a measurement. These three disadvantages arise from the necessity to divide the measurement interval by the same amount that the input is divided.

12.5.5 Heterodyne Converter Counter

The heterodyne converter counter mixes the input signal with a second frequency, and the difference frequency is then counted. The second frequency is usually derived from the counter reference oscillator, which drives a harmonic generator. The desired harmonic is then selected by a tuned filter or cavity, as shown in Fig. 12.20.

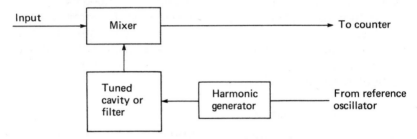

Figure 12.20 Heterodyne converter converts a high input frequency to a lower frequency more suitable for counting.

Heterodyne counters may provide a resolution of 1 Hz, and they operate from dc up to 18 GHz. Their drawback comes from the operator having to use a conversion chart to determine the true input frequency from the readout frequency.

12.5.6 Manual Transfer Oscillator Counter

The manual transfer oscillator counter (TO) is somewhat like the heterodyne converter counter in that two signals are compared. But unlike the heterodyne unit with its fixed reference frequency, the transfer oscillator counter uses a variable frequency oscillator (VFO). The VFO frequency is harmonically related to the input signal being measured. In operation, the counter VFO frequency is counted rather than the input signal's frequency. The correct VFO harmonic is determined by tuning the VFO and noting a zero beat in the meter stage, shown in Fig. 12.21. The counter readout is multiplied by the correct harmonic number to indicate the input signal frequency.

Two advantages of the transfer oscillator counter are a very wide fre-

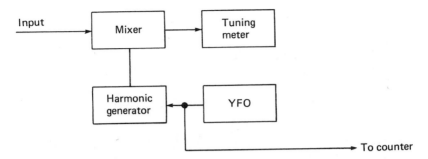

Figure 12.21 Manual transfer oscillator counter measures its own oscillator frequency, which is harmonically related to the input signal.

quency range of 20 Hz to 40 GHz and the ability to measure pulsed signals as well as CW. Disadvantages include loss of resolution, longer measurement time, and the need for operator skill in the complex computation necessary for the harmonic calculation.

12.6 SUMMARY

Signal generating devices include function generators, oscillators, and pulse generators. Although frequency synthesizers have not been discussed in this chapter, they, too, are signal generating devices. A frequency synthesizer is an instrument whose prime function is the generation of any stable precision frequency chosen from a band of frequencies, often many decades in extent. It combines the advantages of a wide-band continuously adjustable oscillator with the frequency precision, at any setting, of a crystal-controlled single-frequency oscillator or frequency standard. Emphasis on each device has been placed on simple theory of operation. Commercial equipment was also presented.

Electronic counters come in many sizes, shapes, and capabilities; some perform many functions, some few. The basic principles of operation, however, are pretty much the same.

The accuracy of electronic counters depends on trigger errors, time base stability, and inherent count ambiguity.

Electronic counter techniques are useful in digital input/output measurements; the coding of digital magnetic tape signals; the measurement of frequency, including 60-Hz line frequency (electronic lab outlets) and signal generator dial frequencies; broadcasting, which includes measurement of the carrier frequency in audio-modulated and frequency-modulated circuitry; and many other electronic measurements.

REVIEW QUESTIONS

1. What is a function generator?
2. What is a test oscillator?
3. What is a signal generator?
4. Draw a block diagram of a simplified FM receiver.
5. How is an FM receiver tested?
6. Describe the IF filter of discriminator alignment procedure for testing receivers.
7. How is a receiver response checked?
8. What is a pulse generator?
9. Describe six ideal pulse generator parameters.
10. Why should the terms rise time and fall time be avoided?
11. Describe the seven actual pulse generator parameters.
12. List and describe the three basic functional blocks of any pulse generator.
13. Why do pulse generators use a 50-Ω source impedance?
14. List five electronic applications of pulse generators.
15. What is the use of an electronic counter?
16. Draw a diagram of the five building blocks of an electronic counter.
17. Describe the seven modes of an electronic counter.
18. What is the function of a heterodyne converter counter?
19. What is the function of a manual transfer oscillator counter?
20. Describe three measurements that can be made with electronic counters.

REFERENCES

1. *TDR for Cable Testing, TDR Cable Application Note 25M1.0.* Tektronix, Inc., Beaverton, Ore, 1976.
2. P. KANTROWITZ, "Fault Location on Telephone Cables," *Trans. Professional Group Commun. Sys. C5-6* (Dec. 1958).

13
RADIO AND MICROWAVE FREQUENCY SYSTEMS

13.1 INTRODUCTION*

In the field of electronic communications, the generation of radio frequency signals serves two principal functions:

1. To act as a carrier for the transmission of sound or other intelligence.
2. To substitute for or simulate such a carrier for the purpose of testing communications devices for performance.

The minimum frequency range requirements for testing various classes of receivers are

Broadcast receivers:	Amplitude modulation, 540–1,605 kHz;
	Frequency modulation, 88–108.5 MHz
Television receivers:	52–216 MHz
Aircraft:	118–136 MHz
Very-high-frequency receivers:	216–990 MHz

All communications systems over 1,000 MHz are considered to be in the microwave frequency range.

In this chapter we shall stress spectrum analyzer measurements, modulation measurements, intermodulation (third-order intercept) distortion

---

*Sections 13.1 through 13.5 are in part based on *Tele-Communication Engineers Instrument Manual*, Caldwell Clements, Inc., New York, 1947.

measurements used in communications systems, receiver alignment procedures, microwave measuring devices, and Citizens Radio Service.

13.2 RADIO FREQUENCY SPECTRUM ANALYZERS

The universal instrument capable of performing radio frequency measurements for a frequency-modulated system, intermodulation distortion, receiver alignment, etc., is the spectrum analyzer. The spectrum analyzer is once more introduced to amplify its use especially in performing radio frequency receiver measurements, which are impossible to accomplish without this instrument.

The radio frequency spectrum analyzer [1, 2, 3, 4] is a device which displays radio frequency signals over a selected portion of the radio frequency spectrum. The display consists of vertical pulses distributed along the horizontal axis of a cathode ray tube screen, the position of each pulse indicating the frequency of a particular radio frequency signal (usually with respect to some *zero reference* frequency at the center of the screen), while the relative height of each pulse indicates the relative strength of each incoming radio frequency signal.

Measurements that can be made with spectrum analyzers include

    1.   Measurements of the continuous-wave oscillator signal (i.e., sine wave).

    2.   One important oscillator signal measurement is its spectral purity. Side bands can result from power supply ripple which are detected by use of a spectrum analyzer. See Figs. 13.1(a), and (b).

    3.   The spectrum analyzer is also suitable for frequency conversion measurements such as the output of a balanced mixer.

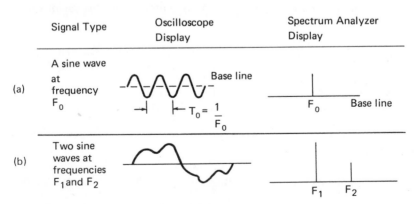

Figure 13.1   Comparison of oscilloscope and spectrum analyzer displays. Courtesy of Tektronix Inc., Beaverton, Oregon.

4. Amplitude modulation measurements including distortion can be made with a spectrum analyzer. See Fig. 13.2(b) for a spectrum of an AM signal.

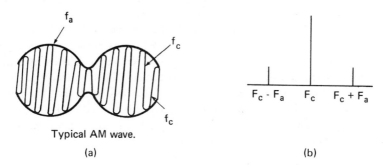

Typical AM wave.

(a)   (b)

Figure 13.2   (a) Typical AM wave. (b) Frequency spectrum of an AM wave.

5. Frequency modulation information such as modulation index and frequency deviation transmitted can be thoroughly given by the spectrum analyzer.

6. By viewing the spectra of a repetitive pulsed radio frequency on the spectrum analyzer, pulse width, average and peak power, frequency, bandwidth employed, and duty cycle can be found. This provides pulsed radio frequency signal information.

7. Microwave measurements such as frequency, bandwidth, SWR, insertion loss, etc., can be accomplished with a spectrum analyzer, it having a frequency capability of thousands of megahertz.

8. Communication circuitry measurements include frequency converters, devices which alter the frequency characteristics of the input signal. Common frequency devices utilized in communications are multipliers, detectors, mixers, and modulators. Frequency multipliers generate higher frequencies by harmonic multiplication and are used to extend the frequency range of oscillators and can achieve large frequency deviations in frequency modulation systems. The mixer available in heterodyne receivers is used when two frequencies are combined to produce another frequency far removed from either of the original frequencies. The modulator places low-frequency information side bands on a high-frequency carrier. The spectrum analyzer can make conversion loss or gain, isolation, distortion, and, in the case of modulators, modulation percentage on these devices.

9. In analyzing digital circuits, pulse rate, pulse width, rise time, overshoot, droop, frequency stability, and transmission bandwidth can be made with a spectrum analyzer.

10. The spectrum analyzer may be used to locate and classify radio transmission in electronic counter measure applications.

13.3 MODULATION MEASUREMENTS

Audio signals by themselves cannot be propagated into space as an airwave and travel great distances. If an attempt is made to transmit audio signals, a transmitting and receiving antenna approximately 2 mi high is required. Signals far above the audio range could be propagated easily into space in the form of electromagnetic waves referred to as radio frequency (RF) waves. These RF waves, called carrier waves, are therefore used as a means of transporting the audio signals to the receivers. This is accomplished by superimposing the audio intelligence on the RF wave such that the new RF wave has one of its characteristics varied in accordance with the audio signal. This superimposition of one signal on another is called *modulation*, with the new wave referred to as a modulated wave.

13.3.1   Amplitude Modulation

One of the first methods of electronic communications was to modulate a carrier wave by turning the carrier signal on and off, transmitting a Morse code. A radio signal of this type is called a *continuous-wave* (CW) signal and is still used in radio communications.

A second and extensively used form of modulation utilized in transmitting audio signals is amplitude modulation. Amplitude modulation (AM) is defined as the variation of the carrier signal amplitude in accordance with the audio signal. See Fig. 13.2(a). The resultant AM wave is transmitted as an air wave, received at the receiving antenna where the AM signal is demodulated, amplified, and converted into intelligence by driving a speaker.

It can be shown mathematically that the AM wave of a single audio tone consists of three separate RF signals of constant amplitude referred to as the carrier, upper side band, and lower side band. If the carrier frequency is noted as f_c and the audio tone as f_a, the upper side-band frequency is $f_c + f_a$, with the lower side-band frequency equal to $f_c - f_a$. See Fig. 13.2(b). In an AM wave each side band contains the entire intelligence signal, with the carrier containing no audio information. For a proper AM signal, the maximum that an audio signal (E_a) amplitude can be is equal to the carrier frequency amplitude E_c, and when this occurs the wave is said to be 100% amplitude-modulated. An audio signal greater than the carrier amplitude results in an overmodulated signal, with the resultant audio signal being highly distorted.

13.3.1.1 Measurement of percent amplitude modulation [5, 6]. In a broadcasting system it is the duty of the transmitter station personnel to see that the average modulation level of the radio frequency carrier is such as to give the best average signal reception over the transmitter serivce area and also to see that this modulation level does not at any time exceed the limits commonly known as "100% modulation." This may involve a rather delicate control of incoming audio signals which are to be used for modulating the transmitter.

Percentage amplitude modulation is defined as

$$m = 100\,\frac{a-b}{a+b}\,\% \tag{13.1}$$

where a is the maximum and b is the minimum amplitude of a modulated wave or *carrier*. See Fig. 13.3. When the radio frequency (RF) carrier is used

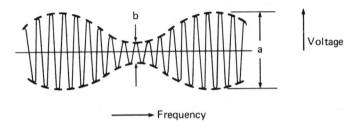

Figure 13.3 Amplitude modulation wave shape with modulation percentage $= m = (a - b)/(a + b)$. (Internal sawtooth sweep of the oscilloscope is applied to the horizontal plates.)

to transmit audio signals by the amplitude modulation method, the average amplitude of the carrier will be represented by $(a + b)/2$ and corresponds closely to the continuous-wave value of the carrier amplitude during periods at no modulation. It is most convenient to measure half of this average peak-to-peak amplitude by means of an RF voltmeter. The audio modulation component may then be isolated by conventional methods of demodulation and measured by a suitable peak-reading voltmeter.

The most convenient method of performing the above measurements is to obtain a sample of the modulated carrier which can be adjusted to a standard level with reference to a calibration mark on the RF or *carrier* voltmeter, whereupon the percent modulation may be read directly from a suitable scale on the voltmeter which measures the modulation component. This method is employed in commercially available modulation monitors, and since the average carrier output level of transmitters is usually quite constant, the input to the carrier level voltmeter will require only occasional standardization.

Any general-purpose oscilloscope may also be used to display the actual modulation envelope of an amplitude-modulated wave. If the modulated carrier frequency is beyond the scope amplifier range, the carrier is heterodyned against a local oscillator of such frequency as to produce an intermediate frequency lying within the range of the oscilloscope vertical amplifier.

See Fig. 13.4(a) regarding utilizing a scope in determining the percent amplitude modulation of a wave. The resulting wave is shown in Fig. 13.4(b).

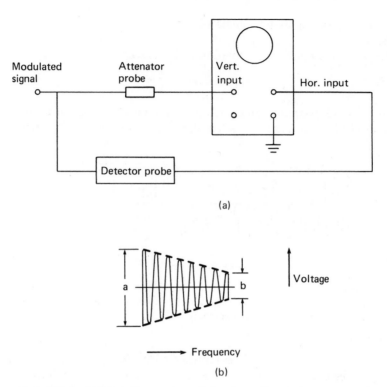

(a)

(b)

Figure 13.4 (a) Block diagram for utilizing the oscilloscope to measure % AM. (b) Oscilloscope amplitude modulation wave shape with modulation percentage $= m = (a - b)/(a + b) \times 100$.

The percent modulation may also be obtained utilizing a spectrum analyzer. See Fig. 13.5 for a 100% amplitude-modulated signal as seen in the display of a spectrum analyzer.

13.3.2 Frequency Modulation

A frequency-modulated system differs from an AM system in that the amplitude of the carrier signal remains constant, while its frequency is altered proportional to the modulation (audio) signal. Two important parameters of the modulation signal must be preserved, namely its amplitude and frequency. The amplitude of the modulation signal is preserved in the FM signal in that the carrier frequency is varied from its unmodulated value at a rate that is directly proportional to the instantaneous modulating voltage amplitude. This change in carrier frequency is called frequency deviation (f_d). The modulation signal's frequency is preserved in that the carrier's frequency (referred to as f_m) is altered at a rate that is directly proportional to the frequency of the modulation signal. See Fig. 13.6 for a carrier signal being frequency-modulated by an audio signal.

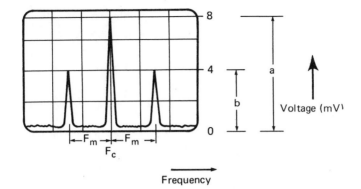

Figure 13.5 Amplitude-modulated signal with 100% modulation as seen in display of a spectrum analyzer. Modulation percentage = m = $(2b/a) \times 100 = 100\%$. (Note: F_c = carrier frequency and F_m = modulating frequency.)

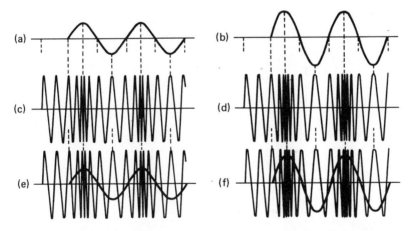

Figure 13.6 The modulating waves in (a) and (b) are the same frequency, but the amplitude of (a) is less than that of (b). The larger signal (b) causes a greater frequency change in the modulated signal at (d), shown by the increased bunching and spreading, than does the smaller amplitude of (a) on the modulated signal at (c). In (e) and (f) the modulating waves are superimposed on their respective modulated waves.

The frequency spectrum of an FM signal differs from an AM signal in that its spectrum is proportional to a modulation index (M_f), defined as

$$\text{modulation index} = M_f = \frac{\text{maximum frequency deviation}}{\text{modulating frequency}} = \frac{f_d}{f_m} \qquad (13.2)$$

See Figs. 13.7(a) and (b) for a typical frequency spectrum of an FM wave. Note that as the modulation index increases, its spectrum bandwidth greatly increases.

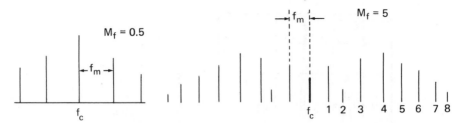

Figure 13.7 RF spectrum of an FM wave as a function of the modulation index, M_f.

Table 13.1
Modulation Index v.
Number of Sideband Pairs

Modulation Index $M_f = \dfrac{f_d}{f_a}$	Number of Side-band Pairs
0.5	2
1	3
2	4
3	6
4	7
5	8
6	9
7	11
8	12
9	13
10	14
11	15
12	16
13	17
14	18
15	19
16	20
17	21
18	23
19	24
20	25
21	26
22	27
23	28
24	29

13.3.2.1 Modulation index of an FM wave measurement.   The simplest and most practical method of determining the modulation characteristics of an FM wave is with a spectrum analyzer. Referring to Fig. 13.7, determine the number of side-band pairs and the distance between side bands. With the amplitude of the side bands decreasing as the spectrum bandwidth increases, a side band is included in the count if its amplitude is at least equal to 1% of the carrier. The distance between side bands is f_m, the modulation frequency.

The modulation index is determined by comparing the number of side-band pairs to Table 13.1 and finding the modulation index. For example with reference to Fig. 13.7(b), the number of side bands is 8, and if f_m is determined as 2 kHz, from Table 13.1, $M_f = 5$, with maximum frequency deviation $f_d = M_f f_m = (5)\,2\text{ kHz} = 10\text{ kHz}$.

13.4   INTERMODULATION (THIRD-ORDER INTERCEPT) DISTORTION

Intermodulation distortion (also known as third-order intermodulation distortion) is one of the most important measurements required in radio and business communications including single side-band transmission.

Intermodulation distortion occurs in the modulation process when harmonics of the fundamental frequency mix with the fundamental and other harmonics. The most troublesome of these is third-order harmonics.

This distortion figure is a problem in amplitude modulation transmission such as in citizens band radios and in frequency modulation transmission such as in two-way radio communication, depending on the bands used.

The following is a typical example of how intermodulation (third-order intercept) distortion occurs. The radio frequency stage of a receiver is highly nonlinear, and feeding into the RF stage is a preferred signal of 1 MHz. Due to the nonlinearity of the radio frequency stage, a second harmonic is created at a frequency of 2 MHz. Also entering the RF stage (which has a wide bandwidth) is a 3-MHz signal from another station. When the 3-MHz signal mixes with the 2-MHz harmonic signal, a resultant frequency of 1 MHz arises. This 1-MHz distortion signal is at the same frequency as our station frequency, causing the term called intermodulation (third-order intercept) distortion. A spectrum analyzer could be used in determining the cause, which is spurious harmonics.

13.5   VISUAL ALIGNMENT SIGNAL GENERATORS FOR RADIO RECEIVERS

13.5.1   Introduction

Visual alignment pertains to a special technique of observing the response frequency characteristics of tuned amplifiers on the screen of a cathode ray oscilloscope. For this purpose, a number of frequency-modulated signal generators have been designed which automatically *sweep* the frequency range of interest. When such a frequency-modulated signal traverses the pass-band of a tuned amplifier to which it is connected, the amplifier output will take the form of an electrical wave whose shape is proportional to the pass-band characteristic of the amplifier. This wave may then be observed on a cathode ray tube screen by synchronizing the start of each signal generator sweep with the start of the oscilloscope time base oscillator sweep. The horizontal, or *time*, axis of the trace then represents the instantaneous frequency of signal generator output, while the vertical axis plots the response characteristic of the tuned amplifier.

13.5.2   Summary of Types

Visual alignment signal generators are differentiated chiefly by their tuning range and sweep width. Of secondary interest to the user is the method of frequency modulation employed, i.e., electronic or electromechanical. General specifications for visual alignment equipment suitable for testing various classes of radio receivers may be summarized as shown in Table 13.2.

Table 13.2
Classes of Radio Receivers

Type of Receiver	Tuning Range (MHz)	Intermediate Frequency	Required Sweep Width
Broadcast "all-wave"	550–30,000	200–500 kHz	10 kHz
Television (low band)	52–88	8.25–26.4 MHz	6 MHz
Television (high band)	174–216	Unknown	Unknown
Frequency mod. (low band)	42–50	4–5 MHz	200 kHz
Frequency mod. (high band)	86–108.5	8–10 MHz	200 kHz

13.5.3   Applications

The advantages of visual alignment are most evident where a quantity of similar band-pass circuits require inspection and adjustment, as in the production testing of radio and television receivers, IF transformers, etc. This is especially true when such circuits must conform to close specifications as to side-band attenuation and uniform sensitivity with the desired pass-band.

For testing FM equipment, as well as the FM sound channels of television receivers, a visual alignment signal generator can perform the dual functions of scanning the RF and IF passband and of generating a suitable frequency-modulated signal for evaluating overall audio fidelity. See Figs. 13.8(a) and (b) for the test setup and resulting scope waveform. Five points of a complete sweep cycle are selected to illustrate successive steps in the development of a visual pattern of band-pass characteristics. The method of coupling the four basic components of the system is also shown schematically.

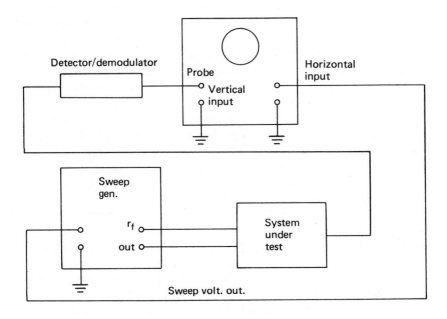

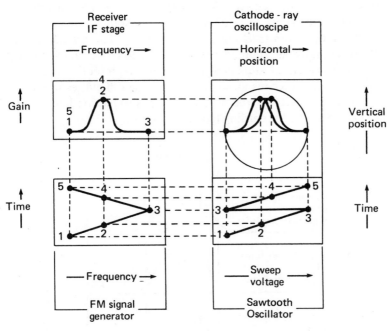

Figure 13.8 (a) Test setup for visual alignment for (b). (b) Four stages in the development of a visual alignment pattern of the double-image type showing the correlation of IF amplifier passband (upper left), pyramid-type FM test signal (lower left), oscilloscope sweep potential (lower right), and the resulting cathode ray tube trace (upper right).

357

In this case, the *pyramid*-type frequency sweep is employed, in which the rates of increase and decrease of signal generator frequency are equal and linear with respect to time. Note that the sawtooth oscillator describes two complete cycles for each sweep cycle of signal generator frequency.

Since the particular tuned stage shown is asymmetrical, two curves appear on the oscilloscope screen, representing the response to the increasing and decreasing frequency of signal generator output. The time intervals 1–2 and 3–4 are not equal; therefore, points 2 and 4 are displaced on the screen. The resulting double image is sometimes useful as an indication of lack of symmetry in the side-band characteristic of tuned circuits.

The three graphs in Fig. 13.9 show three adjustments of a wide-band television IF amplifier as viewed by the *single-image* system of visual alignment. The upper curve shows the flat response (between *B* and *D*) attainable on a good double- or triple-tuned IF stage. If the older point-by-point alignment were attempted on such an amplifier, it would be necessary to tune the signal manually to frequencies *A, B, C, D,* and *E,* the critical alignment points, while at the same time watching a detector output indicator and adjusting trimmers in the receiver. Since television receivers may have 20 or more trimmer adjustments, it is obviously impossible to align 5 points of each circuit for optimum tracking and *flat topping*. The middle and lower curves illustrate improper alignment of double- and triple-tuned stages. It is easy to see how the visual alignment technique, by presenting a continuous picture of tuned amplifier response, overcomes the difficulties of wide-band alignment. This is especially true on the receiver production line, where this technique can be applied according to a fixed routine.

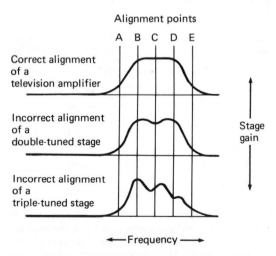

Figure 13.9 Typical response patterns by the visual alignment method.

Alignment of radio receivers (AM, FM) involves use of different IF frequencies and also different band-pass considerations than those of video receivers [7], but the alignment procedures are essentially the same. IF alignment is also discussed in Section 12.3.3.2.

13.6 MICROWAVE MEASURING DEVICES*

13.6.1 Transmission Standing Wave Ratio(SWR) Measurements

For proper operation of a microwave system, the input and output impedance must match the transmission line characteristic impedance (Z_0). Impedance matching with the generator and load assures best power transfer, low interactions of reflected signals and minimizes standing waves. Impedance is a vector quantity, but only amplitude measurements will be considered.

On a microwave transmission line, mismatches of impedance cause reflected waves which interact with the incident wave to form a standing wave pattern. The standing wave ratio (SWR) is the ratio of standing wave maxima to minima along the transmission line and is directly related to impedance mismatch of the component.

The reflection coefficient (ρ) of a device or system is another form of expressing impedance match of devices. The following relationships of ρ and SWR are frequently used in impedance work:

$$\rho = \frac{|E_r|}{|E_i|} = \frac{SWR - 1}{SWR + 1} \quad \text{or} \quad SWR = \frac{E_i + E_r}{E_i - E_r} \tag{13.3}$$

where E_r and E_i are reflected and incident wave E vectors, respectively. Reflection coefficient (ρ) is a linear quantity varying between 0 and 1.0.

See Table 13.3 for typical relations between SWR, reflection coefficient, and return loss in decibels, which is equal to $20 \log_{10} (\rho)$, where ρ is the reflection coefficient.

Table 13.3

SWR	ρ	Return loss
1.0	0.00	α
1.22	0.10	20 dB
2.0	0.33	9.6 dB
α	1.00	0

*Section 13.6 is based on material in the *Coaxial and Waveguide Catalogue and Microwave Measurement Handbook, 1975–1976*, Hewlett-Packard, Palo Alto, California.

There are two general techniques for measuring SWR:

1. Slotted line.
2. Reflectometer and directional bridge.

A slotted line directly detects the ratio of standing wave maxima to minima using a probe moving along the line (SWR). A reflectometer uses directional couplers or bridging techniques to separate incident and reflected waves and measure their ratio (reflection coefficient of return loss in decibels).

The standing wave detectors consist of a slotted section of transmission line, a traveling probe, and associated detecting instruments commonly used for the measurement of impedance at ultrahigh and super-high frequencies. **13.6.1.1 Slotted line techniques—single frequency, coaxial, and waveguide.** The standing wave ratio may be measured directly with a slotted line as shown in Fig. 13.10. The slotted line may be coaxial or waveguide as required

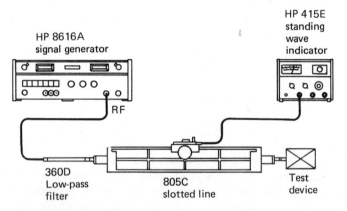

Figure 13.10 Typical setup for SWR measurements with a coaxial slotted line. Courtesy of Hewlett-Packard, Palo Alto, Calif.

by the device under test (DUT). By coupling loosely to the RF field in the line, the probe, when moved along the line, senses the relative field strength of the standing wave pattern, with the SWR read directly on an SWR indicator.

13.6.1.2 Reflectometer techniques for measuring SWR. Reflectometers measure reflection coefficient (ρ) or return loss (dB). The measurement may be made by using two directional couplers. The forward coupler samples the incident signal from the source and monitors its level on the forward detector. The reverse coupler samples the reflected signal from the device under test and senses it at the reverse detector. SWR is calculated as

$$\text{SWR} = \frac{E_i + E_r}{E_i - E_r}$$

The main advantage of the reflector is its broadband swept capability and large dynamic display. For example, adjustments may be made on a device while observing those effects on an oscilloscope display. A typical measurement technique, the coaxial reflector-square law method, will be demonstrated.

A typical coaxial reflectometer measurement system is shown in Fig. 13.11. The HP 11692D dual-directional coupler or equivalent separates

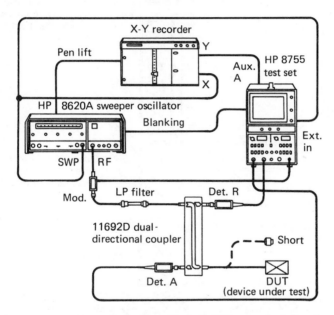

Figure 13.11   Coaxial reflectometer for 2–18 GHz uses dual-directional coupler. Courtesy of Hewlett-Packard, Palo Alto, Calif.

signals into incident (detector *R*) and reflected (detector *A*). Square-wave modulation is imposed on the RF test signal at 27.8 kHz to provide improved sensitivity and lower drift and yet allow fast sweeps.

To calibrate the reflectometer, a short circuit is connected at the output port, thus reflecting all of the incident power. The display instruments are then adjusted for reference to a reflection coefficient of 1.0 (0-dB return loss).

The device under test (DUT) replaces the short, and the actual reflected signal is sampled and displayed. This method depends on the square law detection characteristic of the reflected arm detector. The square law method is generally quite satisfactory for coaxial component work since the range of return loss for typical coaxial components seldom goes beyond 30–35 dB.

The Hewlett-Packard 8755 frequency response test set utilized for detecting the reflected signal and some of its basic operating features are shown in Fig. 13.12. It employs hot carrier diode detectors and provides the appro-

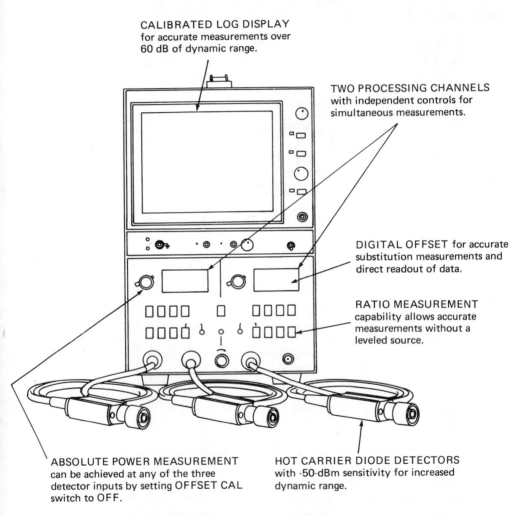

CALIBRATED LOG DISPLAY
for accurate measurements over
60 dB of dynamic range.

TWO PROCESSING CHANNELS
with independent controls for
simultaneous measurements.

DIGITAL OFFSET for accurate
substitution measurements and
direct readout of data.

RATIO MEASUREMENT
capability allows accurate
measurements without a
leveled source.

ABSOLUTE POWER MEASUREMENT
can be achieved at any of the three
detector inputs by setting OFFSET CAL
switch to OFF.

HOT CARRIER DIODE DETECTORS
with -50-dBm sensitivity for increased
dynamic range.

Figure 13.12 HP8755L frequency response test set. Operation with hot carrier diode detectors provides 60 dB of measurement range (+10 to −50 dBm) from 15 MHz to 18 GHz. The system comes complete with an external modulator. Courtesy of Hewlett-Packard, Palo Alto, Calif.

priate linearity compensation for 60 dB of dynamic range over its 15-MHz to 16-GHz operating band.

13.6.2   Attenuation (Insertion Loss) Measurements

Attenuation is a transmission characteristic of a microwave component. It is defined as the ratio of input power to output power for a component measured under perfectly matched conditions (matched source and matched detector).

Actual components have some mismatch and therefore, when measured between "matched" source and detector, yield a power ratio called insertion loss. Insertion loss is higher than the inherent attenuation by the amount of mismatch loss. Insertion loss is of most common interest since typical components have mismatch and exhibit interactions when assembled into larger systems.

Maintaining the detector near its characteristic impedance, Z_0, is accomplished by using well-matched detectors or by putting a well-matched broadband attenuation pad in front of the detector. Ferrite isolators are sometimes used, although they typically give little match improvement over modern detectors. Directional couplers are often used to isolate the detector too, especially in waveguide systems. Pad *isolation* schemes have the disadvantage of losing dynamic measuring range by adding the attenuation of the pad.

Typical detectors are the video LBHCD (low-barrier hot carrier diode), which has a good match and is superior to the older point contact diodes, and the HCD (Schottky) unit, used with the HP 8755L test set (see Fig. 13.12), which has moderate SWR but wider dynamic range due to biasing and shaping. Thermocouple power sensors, having the best inherent SWR, may also be used and are recommended for highest accuracy in attenuation measurement in coaxial, but they also have slow response.

13.6.2.1   Coaxial and waveguide insertion loss measurement—square law method.   The square law (power ratio) technique is probably the most popular one for measuring insertion loss because of its simplicity, low cost, dynamic range, and generally good accuracy. Figure 13.13 shows a typical coaxial measuring setup which uses an HP 11691D, 2- to 10-GHz coupler to provide good source match. During measurement, the reference (R) detector samples incident power to the DUT. The B detector is calibrated for a reference level by placing it directly on the coupler output. After appropriate signal levels and calibration lines are established, the DUT is inserted between source and detector B and the reduction of power in detector B is displayed on a scope or X-Y recorder. Since the test set computes the ratio between detectors R and B, the source need not be leveled.

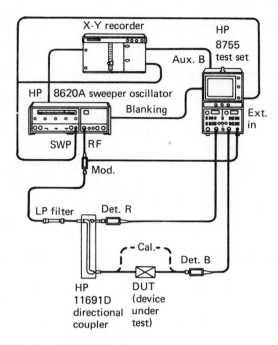

Figure 13.13   Coaxial insertion loss measurement uses a directional coupler for source isolation. Courtesy of Hewlett-Packard, Palo Alto, Calif.

13.7   SUMMARY

Many of the radio frequency instruments have been in existence since 1940. A discussion of the spectrum analyzers, modulation techniques, receiver alignment procedures, and microwave devices used in the radio frequency spectrum has been given. The pulsed radio frequency spectrum analysis has come into much importance in recent studies of pulse spectra or waveforms with lines. A pulse spectrum occurs when the bandwidth of the spectrum analyzer is equal to or greater than the pulse repetition frequency. The spectrum analyzer in this case cannot resolve the actual individual Fourier frequency components since several lines are within its bandwidth. If the bandwidth is narrow compared to the waveform envelope, then the envelope can be resolved. The resultant display is not a true frequency display but a combination of frequency of time and frequency. Such a technique is frequently used in radio frequency pulse waveform analysis.

Emphasis on the reintroduction of the spectrum analyzer and its capabilities has further shown that it is one of the most important tools in performing radio frequency communications measurements. An explanation of AM and FM has been given, with measurement techniques for determining the percent modulation. Receiver visual alignment was also introduced as an important measuring technique for radio-TV servicing. The chapter was concluded with microwave measurements for determining the standing wave ratio and attenuation-insertion loss.

Figure 13.14 gives the electromagnetic spectrum for radio and microwave systems measurements. Table 13.4 gives the names and frequency band for different communications systems.

Table 13.4

Citizen band radio channels	Frequency MHz	Citizen band radio channels	Frequency MHz
1	26.965	21	27.215
2	26.975	22	27.225
3	26.985	23	27.255
4	27.005	24	27.235
5	27.015	25	27.245
6	27.025	26	27.265
7	27.035	27	27.275
8	27.055	28	27.285
9	27.065	29	27.295
10	27.075	30	27.305
11	27.085	31	27.315
12	27.105	32	27.325
13	27.115	33	27.335
14	27.125	34	27.345
15	27.135	35	27.355
16	27.155	36	27.365
17	27.165	37	27.375
18	27.175	38	27.385
19	27.185	39	27.395
20	27.205	40	27.405

There are many other instruments that are utilized in performing radio frequency measurements. Table 13.5 lists some of the popular instruments, several having been described in previous chapters.

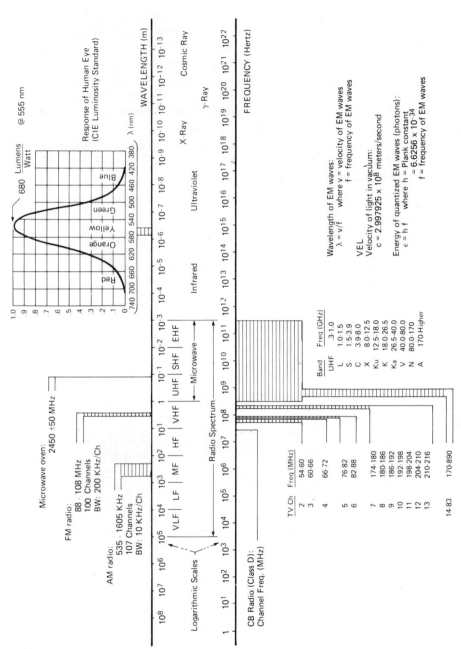

Figure 13.14 Electromagnetic sprectrum.

Table 13.5
Meters Used In Radio Frequency Measurements

Meters	Uses
1. Radio frequency voltmeters	To make radio frequency voltage measurements from 20 kHz to 50 MHz and higher.
2. Q factor meters	To determine figure of merit in an alternating current circuit.
3. Radio frequency bridges	To measure resistance, capacitance, and inductance from which impedance, power factor, Q, etc. can be determined.
4. Intensity meters	Measures the radiated frequency from the broadcasting transmitter for determining radio interference.
5. S meter	Signal strength meter used in sophisticated commercial receivers such as citizen band receivers. The S meter measures the signal strength coming into the receiver at the automatic volume control line or detector output.
6. Grid-dip meter	Measures resonant frequency of a circuit. The meter has several plug-in coils which if placed near another tuned circuit will be loaded down. A capacitor in the circuit is then tuned to a null.

13.8   APPENDIX

WHAT EVERY LICENSEE SHOULD KNOW ABOUT
THE CITIZENS RADIO SERVICE*

Q.   What is the Citizens Radio Service?

A.   The Citizens Radio Service (CB) is a short distance, two-way radio communication service intended for the personal and business use of licensees. The official name for CB is *Class D Citizens Radio Service*.

Q.   Do I need a license in order to use CB radio?

A.   Yes. You need a radio station license issued by the Federal Communications Commission. Operation of a CB transmitter without a valid license will subject the user to a $10,000 fine and/or 1 year imprisonment.

Q.   How can I get a license?

A.   You must complete an application (FCC Form 505) and mail it to the Federal Communications Commission, Gettysburg, Pennsylvania 17326.

Q.   Where can I get the license application (FCC Form 505)?

*Courtesy of the Federal Communications Commission, Safety and Special Radio Services Bureau, Washington, D.C.

A. One is usually packed with every new CB transceiver. Copies of the form may also be obtained from any of the FCC field offices or by writing to the FCC, Room B-10, Attn: Forms Distribution, Washington, D.C. 20554.

Q. *How long does it take to get a CB license?*

A. This varies with work load; however, as of May 1976, it is taking approximately 8 weeks from the time the application is mailed to the FCC in Gettysburg, Pennsylvania before the license is issued. We ask applicants not to inquire into the status of an application before this period has elapsed.

Q. *Who is eligible for a CB radio station license?*

A. Any person 18 years of age or over who has a mailing address in the United States except a person who is a representative of a foreign government.

Q. *Can I obtain a temporary permit to operate my CB station while waiting for issuance of my regular license?*

A. Yes. You must first obtain a copy of FCC Form 555-B entitled *Temporary Permit Class D Citizens Radio Station.* After following the instructions listed on this form you can immediately place your station into operation without any delay. This temporary permit is valid for a period of 60 days from the date you filed your application for a regular license.

Q. *What is meant by a call sign?*

A. Every CB radio station is issued a call sign by the FCC. These call signs currently consist of three letters and four numerals, for example, KYP-2965. A temporary call sign consists of three letters followed by five numerals.

Q. *How must I identify my radio transmissions?*

A. Identify your radio transmissions with your own FCC issued call sign before and after each transmission. "Nicknames" or "handles" may also be used to identify your radio transmissions *provided they are accompanied by the FCC assigned call sign.* It is not necessary to transmit the call sign of the station with whom you are talking.

Q. *How do I go about selecting a channel on which to operate?*

A. In selecting a channel for transmitting, it is very important that the following factors be considered:

1. There are a total of 40 channels available to all CB stations on a *shared basis*;

2. Channel 9 may be used only for emergency communications involving immediate safety of life or protection of property and/or assistance to motorists;

3. You may select any of the remaining 39 channels to conduct your normal personal or business radio communications;

4. To prevent unintentional "blend over" interference to channel 9, the FCC recommends that all transmissions involving highway travelers be conducted on a channel other than channel 10, preferably one several channels removed from channel 9.

Q.   *What are the important factors to consider when using the CB channels?*

A.   Cooperate with all CBers to the fullest extent possible in sharing the CB channels—always try to be courteous and considerate when using a channel. In order to assure that all CB operators will have an equal opportunity to use the frequencies, radio communications between CB stations (interstation) must be limited to no longer than 5 continuous minutes to be followed by a silent period of at least 1 minute. With over 12 million CB operators, there is a need for every CBer to restrict their time on the air to a practical minimum.

Q.   *Who is permitted to operate a CB transceiver?*

A.   If the CB station is licensed in the name of an individual, then the licensee, members of his immediate family residing in his household, and his employees while acting within the scope of their employment may operate the station. If the station is licensed in the name of an organization, then employees or members of the organization may operate the licensee's CB equipment only for performing their duties as employees or members of that organization.

Q.   *What types of communications are permitted to be transmitted in the CB Radio Service?*

A.   Communications to facilitate the personal or business activities of the licensee or communications relating to the following:

1. The immediate safety of life or the immediate protection of property;

2. The rendering of assistance to a motorist, mariner, or other traveler; and

3. Civil defense activities.

Q.   *What is the average "communication range?"*

A.   Since there are many factors that influence the "communication range" (such as terrain, atmospheric conditions, weather, physical obstructions, antenna height, quality of transceiver and antenna, etc.), it is impossible to state with any degree of certainty exactly what the range will be. The average "communication ranges" are as follows:

Mobile to mobile: 5 to 15 miles

Base to mobile: 10 to 25 miles; and

Base to base: 15 to 30 miles.

Q.   Is single-side-band emission type permitted in the CB Radio Service?

A.   Yes. The same FCC license authorizes use of both single-side-band and AM (double-side-band) types of emissions. Both emission types are subject to the same operating rules.

Q.   May a CBer use his CB radio to transmit over long distances (i.e., distances greater than 150 miles)?

A.   No. CBers may not use CB to transmit over 150 miles. Sometimes CB transceivers pick up stations hundreds and even thousands of miles away by means of "skip." Skip occurs when CB radio waves hit the ionosphere and bounce back to earth at least several hundred miles away from their point of origin. Because talking "skip" increases local congestion as well as in the "skip" area, talking via "skip" is prohibited.

Q.   Are codes permitted to be used?

A.   Yes, provided the licensee keeps a list of the codes and their meanings posted with the station license. Various codes in use today are 10-codes, Q signals, and CB jargon, among others. There are no "official" FCC codes.

Q.   What is the best way to select a CB transceiver?

A.   Before you purchase a CB transceiver, make certain it is type accepted by the FCC. Transmitter type acceptance means that the manufacturer has submitted documentary evidence to the FCC which certifies that the transmitter meets the technical standards established by the FCC. Any FCC office or CB equipment dealer may help you determine whether a particular CB transmitter is type accepted. The Commission does not give advice or suggestions regarding the comparative performance of particular brands of radio equipment.

Q.   What procedures must I use to maintain my radio equipment in good order?

A. Have frequency, power, and modulation measurements made at regular intervals. Do not tamper with the equipment. The holder of an FCC Radiotelephone Operator license (First or Second Class) is required for any adjustments that might affect the proper operation of the station.

Q. May I repair my own CB radio transceiver?

A. Yes, but only if you possess a First or Second Class Radiotelephone Operator license or if you do the work under the supervision of the holder of such a license. It is recommended that repairs be made only by a professional technician who has the necessary license and test equipment.

Q. Are power amplifiers permitted to be used by CBers?

A. No. The use of power amplifiers, power boosters, or linear amplifiers is strictly prohibited. Use of these illegal devices can cause serious radio frequency interference problems to TV sets, hi-fi equipment, electronic home entertainment devices, public address (loudspeaker) systems, other CB radio stations, and stations in other radio services. The maximum output power for a conventional, AM, CB transmitter is 4 watts and for single-side-band units 12 watts, peak envelope power.

Q. What are the two CB antenna types, and what are the maximum permissible heights?

A. Two types of antennas used in the Citizens Radio Service are directional and omnidirectional antennas.

1. Directional antennas (e.g., yagi, beam, quad) must be installed so that no part of the antenna exceeds the height of 20 feet above the ground, building, or tree on which it is mounted.

2. Omnidirectional antennas (e.g., whip or ground plane) may be installed so that no part of the antenna exceeds the height of 60 feet above ground level unless the antenna site is near an airport, where a lower antenna height must be used. The antenna may also be installed so that no part exceeds 20 feet above the building or tree on which it is mounted.

If a licensee plans to erect an omnidirectional antenna higher than 20 feet above the ground, he should first review Section 95.37 of the FCC Rules to insure that the installation will not create a hazard to aircraft. CBers may inquire at their local airport, FAA office, or FCC office for assistance in making an aircraft hazard determination. A work sheet (Bulletin 1001h) to assist in determining the maximum permissible height of an antenna structure will be made available from the FCC upon request.

Q.   If I receive a complaint that my CB set is causing radio interference to my neighbor's TV or hi-fi equipment, what should I do?

A.   You should first have your CB equipment checked out by an FCC licensed Radiotelephone Operator (First or Second Class) to make certain your CB equipment meets the FCC technical requirements. If your CB equipment complies with all FCC technical requirements, and if you can demonstrate interference-free reception of TV and other broadcast stations on your own receivers as well as your neighbor's during periods your CB station is in operation, the interference can be assumed to be due to deficiencies in the complainant's receiver and therefore corrective action must be taken by the complainant. A copy of the technician's findings should be kept as part of your station records. After the above procedures, you may suggest to your neighbor that he install a high-pass filter in his TV set.

If additional information is required regarding interference problems, write or telephone the nearest FCC field office. An information bulletin will be sent suggesting further detailed procedures for resolving the interference problem.

Q.   How may I keep informed about the CB Radio Service?

A.   You may obtain the FCC Rules and Regulations (Volume VI, Part 95) governing the CB Radio Service from the Superintendent of Documents, Government Printing Office, Washington, D.C. 20402, for a cost of $1.50. Copies of these Rules are also available at many radio equipment stores at a nominal cost. There are also available from many radio equipment suppliers a variety of excellent publications concerning the CB Radio Service.

Q.   Do I, as a member of the general public, have a voice in the regulation of the CB Radio Service?

A.   Yes. Any person may file a petition to amend or make new Rules. The petition (the original and eleven copies) should be addressed to Federal Communications Commission, Office of the Secretary, Washington, D.C. 20554.

Q.   What can CBers do to assure maximum returns from their CB radio station investments?

A.   CBers should do the following:

1. Voluntarily operate their stations in full compliance with the FCC Rules, and

2. Encourage other CB users to do likewise.

REVIEW QUESTIONS

1. What are the two functions of the generation of a radio frequency signal?
2. Describe how to make a percent modulation measurement.
3. What is the purpose of the radio frequency spectrum analyzer?
4. How does a technician use a visual alignment signal generator?
5. List five spectrum analyzer measurements.
6. What is meant by third-order intercept distortion?
7. What is SWR?
8. What is meant by reflection coefficient?
9. How does a technician make a slotted line measurement in microwaves?
10. How does a technician make an insertion loss measurement in microwaves?

REFERENCES

1. *Spectrum Analyzer Series, Application Notes 150–2, Spectrum Analysis —Pulses RF.* Hewlett-Packard, Palo Alto, Calif., Nov. 1971.
2. *AM, FM Measurements with the Transfer Oscillator, Application Note 141.* Hewlett-Packard, Palo Alto, Calif. Feb. 1, 1974.
3. *The Tracking Generator/Spectrum Analyzer System.* Tektronix, Inc., Beaverton, Ore., Jan. 1976.
4. *The Spectrum Analyzer for Design Engineers, Note No. 5952–0932.* Hewlett-Packard, Palo Alto, Calif.
5. *AM Broadcast Measurements Using the Spectrum Analyzer.* Tektronix, Inc., Beaverton, Ore., Feb. 1976.
6. *An Introduction to Time and Frequency Domain Modulation and Waveform Analysis with Lab Experiments.* Tektronix, Inc., Beaverton, Ore., 1974.
7. WALTER H. BUCHSBAUM, *Television Servicing,* 3nd ed. Prentice-Hall, Englewood Cliffs, N.J., 1975.
8. *High Frequency Swept Measurements, Application Note 183.* Hewlett-Packard, Palo Alto, Calif. Nov. 1975.

 9. *Coaxial and Waveguide Catalogue and Microwave Measurement Handbook, 1975–1976.* Hewlett-Packard, Palo Alto, Calif. 1975.
10. Hewlett-Packard Application Note Index, Hewlett-Packard, Palo Alto, Calif., October 1975.

GLOSSARY
OF
INSTRUMENT TERMS*

accuracy   The capability of an instrument to follow the true value of a given phenomenon. Often confused with *inaccuracy*, which is the departure from the true value into which all causes of error are lumped—including *hysteresis, non-linearity, drift, temperature effect*, etc. While this definition is unique, many authors and speakers use the word *accuracy* as a synonym (incorrectly) for *repeatability*, which is defined elsewhere.

ambient temperature   The prevailing temperature in the immediate vicinity or the temperature of the medium surrounding an object.

amplitude   The magnitude of variation in a changing quantity from its zero value. The word must be modified with an adjective such as "peak," "rms," "maximum," etc., which designates the specific amplitude in question.

amplitude response   The maximum output amplitude obtainable at various points over the frequency range of an instrument operating under rated conditions.

analog   An adjective which has come to mean continuous, cursive, or having an infinite number of connected points. The instrumentation industry uses the words "analog" and "digital" where the more precise language would be "continuous" and "discrete." Short for "analogous."

aperiodic damping   or overdamping. Damping so great as to prevent a normally oscillating body or electrical system from *passing through the rest position* upon removal of the external force.

*From the revised third edition, published Feb. 1973, with permission of Gould, Inc., Instrument Systems Division, Cleveland, Ohio, copyright ®.

attenuation   Reduction or division of signal amplitude—retaining the characteristic waveform; implies deliberately throwing away or discarding a part of the signal energy for the sake of reduced amplitude.

balanced bridge   A Wheatstone bridge circuit, which when in a quiescent state has an output voltage of zero.

balanced input   A symmetrical input circuit having equal impedance from both input terminals to ground.

balanced line   A transmission line consisting of two identical signal conductors having equal resistance and equal capacities with respect to the cable shield or with respect to ground.

bandwidth   The range of frequencies over which a given device is designed to operate within specified limits.

bonded strain gage   Strain-sensitive elements arranged to facilitate bonding to a surface in order to measure applied strain on that surface.

breakdown voltage   The voltage at which electrical equipment will arc over to a structural member. Also voltage at which a component's insulation fails.

calibration

(1) The process of comparing a set of discrete magnitudes or the characteristic curve of a continuously varying magnitude with another set or curve previously established as a standard. Deviation between indicated values and their corresponding standard values constitutes the correction (or calibration curve) for inferring true magnitude from indicated magnitude thereafter.

(2) The process of adjusting an instrument to fix, reduce, or eliminate the deviation discussed in definition (1).

calibration curve

(1) The path or locus of a point which moves so that its coordinates on a graph are corresponding values of input signals and output deflections.

(2) The plot of error versus input (or output).

chopper   An electromechanical or electronic device used to interrupt a dc or low-frequency ac signal at regular intervals to permit amplification of the signal by an ac amplifier.

combined error   A term to denote the largest possible error in an instrument resulting from a combination of adverse conditions. Often applied to the largest error attributable to nonlinearity and hysteresis.

common mode rejection (or in-phase rejection)   A measure of how well a differential amplifier ignores a signal which appears simultaneously and in phase at both input terminals. Usually and preferably stated as a voltage ratio but often stated in the dB equivalent of said ratio at a specified frequency, e.g., "120 dB at 60 Hz with 1-kΩ source impedance."

common mode signal   A signal that appears simultaneously at both amplifier input terminals with respect to a common point.

critical damping   The value of damping which provides the most rapid *step-function response* without overshoot. The borderline between *aperiodic damping* and underdamping.

crosstalk   Interference in a given transmitting or recording channel which has its origin in another channel.

cursive   Connected or continuous. Longhand is a cursive form of marking, whereas printed letters are a noncursive, discrete form of marking. The word "analog" is often taken for a synonym of *cursive*.

D'Arsonval galvanometer   A device in which direct current flowing through a coil pivoted between the poles of a permanent magnet reacts with the latter's magnetic field, causing a torque.

dc amplifier   One which has a frequency response band that goes down to dc. This definition includes direct-coupled amplifiers but is not restricted to them.

damped natural frequency   The natural frequency of a system which has damping in it.

damping   An electrical, mechanical, or magnetic force which opposes oscillation of a body or system capable of free oscillation. Classically, damping is that situation which prevails when the potential and kinetic energies of a vibrating system are unequal, and the difference between the two dissipated as heat.

damping factor

(1) The quotient obtained by dividing the Naperian logarithm of the ratio of two successive excursions by the time interval between them (i.e., the one-cycle time). The excursions are measured by maximum amplitude from rest position or zero deflection. In this definition an undamped system has a damping factor of zero.

(2) Instrument damping factor is the amplitude ratio of any one excursion to the next succeeding excurison in the same direction when energy is not supplied. Under this definition an undamped system has a damping factor of 1.0.

(3) The ratio of the system friction to that amount of friction required to make the system critically damped. Under this definition, an undamped oscillator has a damping factor of zero.

Brush technical literature is based on the third definition, which is compatible with the first in that an undamped system has a damping factor of zero. Note that a damping factor of 1.0 describes critical damping in definition (3).

decibel   abbreviated dB. Ten times the logarithm of the ratio between two amounts of power P_1 and P_2 existing at two points or at two instants in time. By definition,

$$\text{number of dB} = 10 \log_{10} \frac{P_2}{P_1} = 20 \log_{10} \frac{E_2}{E_1} + 10 \log_{10} \frac{Z_1}{Z_2}$$

assuming the power factors of the two impedances are equal. If the impedances themselves are equal, the right-hand term becomes zero and

$$\text{number of dB} = 20 \log_{10} \frac{E_2}{E_1}$$

degrees of freedom   A phrase used in mechanical vibrations to describe the complexity of a system. The number of *dof* is the number of variables which can be fixed at will and the state of a vibrating system defined thereby. One degree of freedom implies a single oscillator (consisting of one mass, one spring, and one damping force), two degrees of freedom imply two oscillators interacting on each other, etc.

demodulator   A device which extracts the modulation information from a modulated carrier. In most cases it rectifies the incoming modulated carrier frequency and separates the desired modulation signal from the carrier.

differential input   An input circuit that rejects voltages which are the same at both input terminals and amplifies the voltage *difference* between the two input terminals. May be either balanced or floating and may also be guarded.

differential transducer   A device which is capable of following simultaneously the voltages across or from two separate signal sources and providing a final output proportional to the difference between the two signals.

digital output   An output quantity consisting of a set of discrete magnitudes coded to represent digits in a system of notation. Most common example of digital output is an electric typewriter, others include "nixie tubes," drum counter, etc.

discrete   Made of distinct parts, usually numerals or other symbols which stand for quantities or single values. An abacus is a simple computer which handles *discrete* information, while a slide rule is a simple computer which handles *analog* information and from which discrete answers may be estimated. The word discrete is always associated with a *finite* number of items in a set; cursive (or "analog") is always associated with an *infinite* number of points in a line or curve.

discriminator   A device in which the properties of a signal such as frequency or phase are converted into amplitude variations.

distortion   An unwanted change in waveform. Principal forms of distortion are inherent nonlinearity of the device, nonuniform response at different frequencies, and lack of constant proportionality between phase shift and frequency. (A wanted or intentional change might be identical, but it would be called *modulation*.)

distributed capacity   Capacitance evenly distributed over the entire length of a signal cable. Includes capacitance between signal conductors and from each conductor to ground.

drain wire   A metallic conductor frequently used in contact with foil-type signal-cable shielding to provide a low-resistance ground return at any point along the shield.

drift   A gradual and unintentional deviation of a given property—in the case of a recording pen usually the deviation from zero rest position.

dynamic error band   The word "band" in this case does not refer to a portion of the frequency spectrum; it means the spread or band of output-amplitude deviation incurred by a *constant-amplitude sine wave* as its frequency is varied over a specified portion of frequency spectrum. The spectacle named by this term is of course the same one named by *frequency response*—except for the restriction on input. Good frequency response would be displayed by faithful throughput of a square wave, a test which, by definition, would be unrelated to dynamic error band. See *static error band*.

dynamic run   The test performed on an instrument to obtain the overall behavior and to establish or corroborate specifications such as frequency response, natural frequency of the device, etc.

electrocardiogram   Essentially an electromyogram of the heart muscle. All muscular activity in the body is characterized by the discharge of polarized cells, the
aggregate current flow from which causes a voltage drop which can be measured
on the skin. A changing EMF will appear between electrodes connected to the
arms, legs, and chest which rises and falls with heart action such that the period
of the resulting waveform is the time between heartbeats. Various positive and
negative peaks within one cycle of this waveform have been lettered P, Q, R, S, and
T, a notation which aids in subsequent analysis and diagnosis.

electrode
   (1) A conductor, by means of which a current passes into or out of a fluid or
an organic material such as human skin; often one terminal of a *lead*.
   (2) The emitter, collector, or base of a semiconductor.
   (3) The cathode, grid, or anode of a vacuum tube or gas-discharge lamp.
   (4) A metallic conductor such as in an electrolytic cell, where conduction by
electrons is changed to conduction by ions or other charged particles.
   (5) A device which combines definitions (1) and (4), e.g., a calomel half-cell.
   It is conceivable that under definition (4) an electrode could be a kind of
transducer.

electroencephalogram   A waveform obtained by plotting brain voltages (available
between two points on the scalp) against time. An electroencephalogram is not
necessarily a periodic function, although it can be—particularly if the patient is
unconscious. These voltages are of extremely low level and require recording
apparatus which displays excellent noise rejection.

electromagnetic   Pertaining to the mutually perpendicular electric and magnetic
fields associated with the movement of electrons through conductors, as in an
electromagnet.

electromyogram **(EMG)**   Classically, a waveform of the contraction of a muscle
as a result of electrical stimulation. Usually said stimulation comes from the
nervous system (normal muscular activity). The record of potential difference
between two points on the surface of the skin resulting from the activity or action
potential of a muscle.

electrostatic coupling   Coupled by means of capacitance so that charges on one
circuit influence another circuit owing to the capacitance between the two.

floating   The condition of a device or circuit that is not grounded and not tied to
any established potential.

floating input   An isolated input circuit not connected to ground at any point (the
maximum permissible voltage to ground is limited by electrical design parameters
of the circuit involved). It is understood that in a floating input circuit both conductors are *equally* free from any reference potential, a qualification which limits
the types of signal sources which can be operated floating.

force-balance transducer   A transducer in which the output from the sensing member is amplified and fed back to an element which causes a force-summing member to return to its rest position. The magnitude of the signal fed back determines
the output of the device like the error signal in a servo system.

frequency-modulated signal   A signal where the intelligence is contained in the deviation from a center frequency. In a recording instrument, this deviation is proportional to the applied stimulus.

frequency response   The portion of the frequency spectrum which can be covered by a device within specified limits of amplitude error.

full scale   The total interval over which an instrument is intended to operate. Also the output from a transducer when the maximum rated stimulus is applied to the input.

gain stability   The extent to which the sensitivity of an instrument remains constant with time. The property reported in specifications should be *instability*, which is the maximum change in sensitivity from initial value over a stated period of time under stated conditions.

ground

   (1) A point in a circuit used as a common reference or datum point in measuring voltages.

   (2) The conducting chassis or framework on which an electrical circuit is physically mounted and to which one point in a circuit is often connected.

   (3) The earth or a low-resistance conductor connected to the earth at some point and having no potential difference from another conductor connected to the earth at the same point.

   Please remember that the first definition doesn't necessarily have anything to do with the other two.

ground loop   The generation of undesirable signals within a ground conductor, owing to circulation currents within the conductor which originate from a second source of potential—frequently as a result of connecting two separate grounds to a signal circuit.

guard shield or "guard"   An internal floating shield which surrounds the entire input section of an amplifier. Effective shielding is achieved only when the absolute potential of the guard is stabilized with respect to the incoming signal.

guarded input   An input that has a third terminal which is maintained at a potential near the input-terminal potential for a single-ended input—or near the mean input potential for differential input. It is used to shield the entire input circuit.

hysteresis   The summation of all effects, under constant environmental conditions, which causes the output of an instrument to assume different values at a given stimulus point when that point is approached first with increasing stimulus and then with decreasing stimulus. *Hysteresis* specifically includes backlash. The word is most typically applied to the relationship between magnetizing force and magnetic flux density in an iron core transformer. In the instrumentation field, *hysteresis* is the same thing as dead-band.

impedance   An indication of the total opposition that a circuit or device offers to the flow of alternating current at a particular frequency. A combination of resistance R and reactance X (at a designated frequency)—all expressed in ohms.

$$|Z| = (R^2 + X^2)^{1/2}$$

inductive pickup   Signals generated in a circuit or conductor owing to mutual inductance between it and a disturbing source.

infinite resolution   A phrase used as a synonym for cursive, continuous, or analog. Capable of stepless adjustment. Has nothing to do with *resolution*, really.

insulated   A condition when the substance between one conductor, circuit, or device and another cannot conduct current at operating voltage and beyond (to the limit of breakdown voltage). The substance could be air as in the case of a bus bar, liquid (e.g., oil) as in a power transformer, or a solid as in the substrate of a printed-circuit board. See *isolated*.

internal calibration   Calibration by an internal voltage source (provided with the instrument) rather than an external standard.

isolated   Utterly cut off from; refers to that condition where a conductor, circuit, or device is not only *insulated* from another (or others), but the two are mutually unable to engender current, EMF, or magnetic flux in each other. As commonly used, *insulation* is associated predominately with dc, whereas *isolation* implies additionally a bulwark against ac fields.

lead

(1) An electrical conductor or wire used to connect a test point to an electronic device or instrument; a wire terminated by an electrode.

(2) A particular connection scheme for the input leads in an electrocardiogram e.g., voltage between right arm and left arm constitutes "lead I." Number of leads taken during an electrocardiogram may vary between 8 and 12, normally designated I, II, III, AVR, AVL, AVF, V, CR, CL, and CF (first three are the most significant).

(3) The particular oscillogram resulting from one of the connections above.

linearity   The "straight-lineness" of the transfer curve between an input and an output; that condition prevailing when output is directly proportional to input. See *nonlinearity*.

loading error   A loss of output signal from a device owing to a current drawn from its output. It increases the voltage drop across the internal impedance where we would really like to have no voltage drop at all.

log/linear preamplifier   A preamplifier whose electrical output is proportional to the common logarithm of the input but which can be switched to a different mode wherein output is proportional to input. Any amplifier whose transfer characteristic or calibration curve can be arbitrarily switched from the form $y = \log x$ to the form $y = x$ (y equals output and x equals input).

low-level signal   Very small amplitudes serving to convey information or other intelligence. Variations in signal amplitude are frequently expressed in microvolts.

low-pass filter   A filter that transmits alternating current below a given cutoff frequency and substantially attenuates all other currents.

magnetic damping   The damping of a mechanical motion by means of the reaction between a magnetic field and the current generated in a conductor moving through that field. The resistance of this conductor converts excess kinetic energy to heat (see *damping*).

magnetic field   Any space or region in which magnetizing forces are of significant magnitude with respect to conditions under consideration. A magnetic field is produced by any current flow or a permanent magnet including the earth itself.

marking   Impressing or causing to appear on some surface a visible trace, line, dot, or symbol. Handwriting is one form of cursive marking.

natural frequency   The frequency at which a system with a single degree of freedom will oscillate upon momentary displacement from rest position by a transient force in the absence of damping.

noise   Any unwanted electrical disturbance or spurious signal which modifies the transmission, measurement, or recording of desired data.

nonlinearity   The prevailing condition (and the extent of its measurement) under which the input/output relationship fails to be a straight line. This relationship is known as the input/output curve, transfer characteristic, calibration curve, or response curve. Nonlinearity is measured and reported in several ways, and it is necessary to state *which* way along with the magnitude in any specification.

(1) *Maximum-deviation-based nonlinearity* is the maximum departure between the calibration curve and a straight line drawn to give the most favorable accuracy. Normally expressed as a percent of full-scale deflection.

(2) *Slope-based nonlinearity* is the ratio of maximum slope error anywhere on the calibration curve to the slope of the nominal sensitivity line. Usually expressed as a percent of nominal slope.

Nearly all of the many variations beyond these two difinitions result from the many ways in which the straight line can be arbitrarily drawn. All are valid as long as construction of the straight line is explicit.

normal mode   The expected or usual operating conditions, such as the voltage which occurs between the two input terminals of an amplifier.

notch filter   An arrangement of electronic components designed to attenuate or reject a specific frequency band with sharp cutoff at either end.

null balance   A condition of balance in a device or system which results in zero output or zero current input.

off-ground   The voltage above or below ground at which a device is operated.

operating temperature   The temperature or range of temperatures over which an instrument is expected to operate within specified limits of error.

overshoot   A condition whereby the initial response or deflection is momentarily greater than it ought to be for the new steady state called for by an abrupt change in signal. The little blip on top of a square wave at the ascending corner.

parameter   Literally, a measuring device that sits beside something—para(beside) + meter (a device for measuring), as in a coefficient which sits beside the variable it multiplies. A parameter is either a dimensionless constant or a function of time, in either case derived from the mathematical relationship between a group of *properties* whose values are fixed. It then takes the place of these properties and represents them as one family in subsequent physical relationships. (For the difference between "property" and "parameter," see *property*.) The several parameters of a physical system (such as an oscillograph) are sufficient to determine its response from any stimulus.

pen centering   An electrical or mechanical adjustment by which an oscillograph pen is positioned to channel center.

pen positioning   An electrical or mechanical adjustment by which an oscillograph pen is positioned to any desired amplitude grid mark on the chart to represent zero signal.

period   The time required for a complete oscillation or for a single cycle of events. The reciprocal of *frequency*.

periodic damping   Less-than-critical damping; the kind of damping present in an underdamped system.

phase   In a periodic function or wave, the fraction of the period which has elapsed measured from some fixed origin. If the time for one period is represented as 360° along a time axis, the phase position is called the *phase angle*.

phase angle
   (1) The angle between two vectors representing two simple periodic quantities which vary sinusoidally and which have the same frequency.
   (2) A notation for phase position when the period is designated by 360°.

phase shift   A change in the phase relationship between two periodic functions. See *phase*.

phonocardiogram   A graphic recording of the sounds produced by the heart and its associated parts, e.g., its mitral and aortic valves.

precision   A word so closely related to resolution as to be nearly synonymous. Sharpness or clarity. The smallest part that a system or device can distinguish. Precision is a word associated with *possible* or *design* performance, whereas accuracy and repeatability reflect *actual* performance. For example, a micrometer calibrated to the nearest 10^{-4} in. is more precise than one calibrated only to the nearest 10^{-4} ins. Both micrometers could be equally inaccurate.

primary calibration   A calibration procedure in which the instrument output is observed and recorded while the input stimulus is applied under precise conditions—usually from a primary external standard traceable directly to the U.S. Bureau of Standards.

property   Properties are the physical quantities which directly describe the physical attributes of the system (whereas parameters are those combinations of the properties which suffice to determine the response of the system). For an oscillograph, maximum sensitivity, input impedance, maximum output voltage, and most of the other items listed in its table of specifications are *properties*. Frequency response and rise time are generally parameters. Properties have all sorts of dimensions; a parameter is usually dimensionless but may have the dimension of time. See *parameter*.

QRS *complex*   That portion of the waveform in an electrocardiogram extending from point Q to point S, it includes the maximum amplitude shown in an ECG trace.

range   A statement of the upper and lower limits between which an instrument's input may be received and for which the instrument is calibrated.

rated range   The nominal operating range within which a device can be operated and still maintain the performance characteristics specified by its manufacturer.

reactive balance

(1) The condition of an ac circuit where the phase angle between voltage and current is zero.

(2) The amount of corrective capacitance or inductance required to null the output of certain transducers or systems having ac excitation.

reliability   A statistical term having to do with the *probability* that an instrument's repeatability and accuracy will continue to fall within specified limits. The attribute which makes an instrument work day after day the way it is supposed to work.

repeatability

(1) The maximum deviation from the mean of corresponding data points taken from repeated tests under identical conditions.

(2) The maximum difference in output for any given identically repeated stimulus with no change in other test conditions.

Many people use the word *accuracy* when they mean repeatability.

resistance balance   The amount of resistance which is required to null the output of certain transducers or input systems.

resistance strain gage   A metallic wire or foil grid that produces a resistance change directly proportional to its change in length (strain). Its output is often calibrated in terms of other quantities which vary with strain.

resolution   The smallest change in applied stimulus that will produce a detectable change in the instrument output. Resolution differs from *precision* in that it is a psychophysical term referring to the smallest increment of *humanly perceptible* output (rated in terms of the corresponding increment of input).

resonant frequency   The first point in the frequency spectrum (other than dc) where a device looks purely resistive or where all energy supplied is used to overcome friction and is dissipated as heat. When no damping is present the resonant frequency and the undamped natural frequency are identical. When damping is present the resonant frequency is always *lower* than the undamped natural frequency.

ringing   Plucking a guitar string, striking a piano string, or twanging a spring-restored pen could be examples of ringing: the continued oscillation of a vibrating system after the removal of external force (including voltage).

scale factor

(1) The amount by which a quantity being measured must change in order to produce unit pen deflection.

(2) The ratio of real to analog values.

sensitivity   The property of an instrument which determines scale factor. As commonly used, the word is often short for "maximum sensitivity," or the minimum scale factor with which an instrument is capable of responding.

sensitivity inaccuracy   The maximum error in sensitivity displayed by a pen as a result of the summation of all of the following: *frequency response, attenuator inaccuracy, hysteresis* or deadband, *amplitude distortion* (*senstivity nonuniformity*), *phase distortion* (change in phase relationship between input signal and output deflection), and *gain instability.* Only by taking into account all six of these factors can nominal sensitivity as indicated by the numeral on the attenuator dial be properly discounted for accurate interpretation. The parameter of sensitivity inaccuracy is reported in most Brush specifications.

sensitivity nonuniformity The nonuniformity of the input/output ratio across the scale of an instrument. Derived from observing the nonuniformity of output increments corresponding to a succession of uniform input increments (say, 10% full range) imposed on different parts of the scale. As a practical matter, the specification so derived includes hysteresis and stiction. In this respect it differs from the abstract concept of *slope-based nonlinearity*, which does not include these errors. If the worst condition encountered in the above succession were, say, 10.3% of full scale for an output increment that averages 10.0%, then the *sensitivity nonuniformity* of the device would be $(10.3 - 10) \div 10 = 0.03 = 3\%$.

shunt calibration A form of secondary calibration in which a resistor is placed in parallel across one element of a resistive bridge in order to obtain a known and deliberate electrical unbalance.

shunting effect A reduction in signal amplitude caused by the load which an amplifier or measuring instrument imposes on the signal source. For dc signals the shunting effect is directly proportional to the output impedance of the signal source and inversely proportional to the amplifier's input impedance.

signal conditioner A device placed between the signal source and the pendrive amplifier to condition the signal. Examples: damping networks, attenuator networks, preamplifiers, excitation and demodulation circuitry, signal converters (for changing one electrical quantity into another, e.g., volts to amps), instrument transformers, equalizing or matching networks, filters, etc.

single-ended (amplifier) An amplifier with one input terminal and one output terminal tied to a common point and therefore operating at a common potential. This point may or may not be connected to *ground*. The phrase "single-ended input" does not refer simply to a connection scheme but refers explicitly to the input of a *single-ended amplifier*.

single-point grounding A grounding system that attempts to confine all return currents to a network which serves as the circuit reference. The phrase *single-point grounding* does *not* imply that the grounding system is limited to one earth connection. To be effective, no appreciable current is allowed to flow in the circuit reference; i.e., the sum of above return currents is zero.

span The reach or spread between two established limits such as the difference between high and low values in a given range of physical measurement.

stability
(1) Independence or freedom from changes in one quantity as the result of a change in another.
(2) The absence of drift.

static error band
(1) The spread of error that would be present if the pen were stopped at some deflection—say, at half-scale. Like nonlinearity, its value depends on pen position and is reported as a percent of full scale.

(2) A specification or rating of maximum departure from the point where the pen ought to be when an on-scale signal is stopped and held at a given signal level. This definition explicitly stipulates the stopped position may be approached from either direction in the course of following any random waveform. Therefore, it is a quantity which includes hysteresis and nonlinearity but does not include chart-paper accuracy or electrical drift. The term is comparable to *dynamic error band*.

step function response   Or rise time. The characteristic curve or output plotted against time resulting from the input application of a *step function* (a function which is zero for all values of time prior to a certain instant and a constant for all values of time thereafter).

stiction   Friction, or the "sticking" effect during that infinitesimal time between rest and finite velocity. In Brush apparatus, it is displayed primarily as the result of friction between pen tip and chart paper.

strain   The physical deformation, deflection, or change in length resulting from stress (force per unit area). The magnitude of strain is normally expressed in microinches per inch.

strain-gage-based   An instrument or transducer whose sensing element is composed of bonded or unbonded strain gages.

temperature effect   The change in performance owing to changes in temperature. It is specifically the difference between output at room temperature and at some other specified temperature from the same stimulus value when all other conditions are constant. It is generally specified as a percentage of full-scale output per interval of temperature.

thermal coefficient of sensitivity   The change in fullscale output caused by temperature effect. The phrase is simply a statement of the *temperature effect* for a specific instrument.

thermal zero shift   The maximum change or shift in pen zero resulting from changes in temperature. Simply a statement of the *temperature effect* for a specific instrument at *zero signal*.

thermistor   A semiconductor whose resistance is extremely temperature-sensitive. Like carbon (formerly alone among conductors), thermistors have negative temperature coefficients. Thermistors are used to compensate for temperature variations in other parts of a circuit, and they are also used as transducers.

thermocouple   A temperature-sensing device consisting of two dissimilar metal wires joined together at both ends to deliberately incur the Seebeck effect. One wire of the circuit is opened (both output terminals are the same metal), and a small EMF appears across the terminals upon application of heat at one junction. The magnitude of this EMF is proportional to the *difference* in temperature between the "measuring junction" (located at the point of measurement) and the "reference junction" (usually located in the measuring instrument or in a tumbler of icewater). Special alloys have been developed to serve as the dissimilar metals, i.e., constantan, alumel, chromel, and platinum-rhodium. These alloys are paired with each other or with pure-element metals such as copper and iron.

threshold of sensitivity   The smallest stimulus or signal that will result in a detectable output. This phrase is frequently used to describe the voltage point at which an operations monitor or event marker will trigger.

time base
(1) A reference time signal recorded at given intervals with the information signal.
(2) The axis of chart motion.
(3) The abscissa of many plotted curves.

time constant
(1) The time required for an exponential quality to change by an amount equal to 0.632 times the total change required to reach steady state.
(2) The number of seconds required for a capacitor in an RC circuit to reach 63.2% of full charge after the application of a step function voltage.

transducer   A device for translating faithfully the changing magnitude of one kind of quantity into corresponding changes of another kind of quantity. The second quantity often has dimensions different from the first and serves as the source of a useful signal. It is convenient to think of the first kind of quantity as an "input" and the second kind as an "output." Thus the input quantity to a dynamic microphone is air *pressure*; the output quantity is *voltage*.

There may or may not be significant energy transfer from the transducer's input to output. For example, a photocell transducer does transfer energy; a strain-gage-based transducer does not.

transient   A sudden signal change of short duration characterized by a steep wavefront which, when plotted against time, would appear markedly different from the waveforms immediately preceding and following it. "Step function" is the term properly used in connection with establishing ratings.

transient response   A phrase whose correct use is rare but possible. Chances are you mean "step function response." Good step function response will provide faithful recording of a transient.

tuned filter   An arrangement of electronic components which either attenuates signals at a particular frequency and passes signals at other frequencies or vice versa.

turns ratio   The ratio of the number of turns in one winding of a transformer to the number of turns in another winding.

vernier   Any device, control, or scale used to obtain fine adjustment or more accurate measurement than the main measuring scale.

voice frequency   The frequency range of ordinary speech usually considered to be a band from 100 Hz to 3 kHz.

volt-ampere   A unit of apparent power equal to the product of volts times amperes without taking into account the power factor. Volt-amperes times power factor equals power in watts.

watt   The practical unit for power and therefore the *rate* at which energy is converted to work or dissipated as heat. In the case of electricity, power P in watts equals the product of voltage times *in-phase* current. For dc, $P = EI$; for ac, $P = EI \times$ power factor. See *volt-ampere* above.

waveform   The graphical representation of a wave, formed by plotting amplitude versus time. Wave *shape*.

Zener diode   A semiconductor used as a constant voltage reference or control element in various electronic circuits—particularly power supplies. Differs from other diodes in that its electrical properties are derived from a rectifying junction which works at a reverse bias avalanche condition.

zero adjustment   Bringing the pointer or pen of an instrument to zero when the input signal is zero.

zero-line stability   Absence of drift when the instrument is at pen zero.

zero suppression   A technique of bucking out part of an incoming signal (the static component) so that the rest of a signal (the dynamic component) may be amplified and displayed with greater expansion on a recorder chart. Calibrated zero-suppression is a more elaborate, precalibrated system which clearly indicates the magnitude of the static component which is being bucked out for any given setting of the zero suppression control. The zero-suppression control is entirely different from pen-position control. By removing the static component *prior* to attenuation, zero suppression allows the full gain capability to be applied to the dynamic portion of the signal. Pen-position signal, following *after* attenuation, can only buck the *attenuated* static component and thus wastes a part of the gain capability which could otherwise be applied to the dynamic component (as it is with zero suppression).

Table A-2-1
dBm to Voltage Conversion

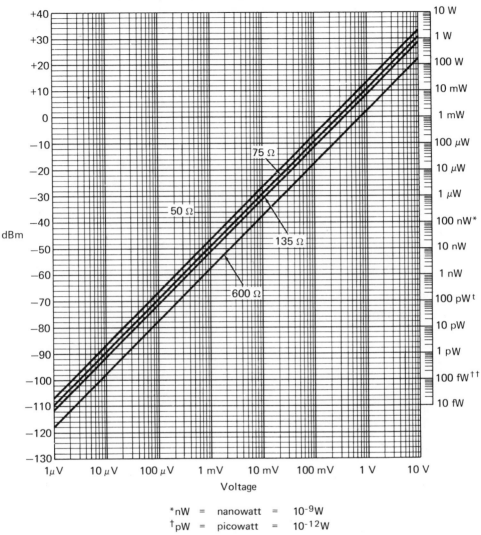

Voltage

*nW = nanowatt = 10^{-9}W

†pW = picowatt = 10^{-12}W

††fW = femtowatt = 10^{-15}W

To convert dBm to voltage, move to the right from the dBm scale to the line representing the impedance of the circuit being measured. Then move down and read the signal strength on the voltage scale. Use the reverse procedure to convert voltage to dBm.

A power scale has also been provided to allow conversion from voltage to power or dBm to power. Note that the relationship between power and dBm does not depend on circuit impedance.

For example, −30 dBm (which is 1 μW for all impedances) is 7 mV for a 50 Ω impedance, but it is 24 mV for a 600 Ω impedance.

Courtesy of the copyright holder, Hewlett-Packard, Palo Alto, Calif.

INDEX